Advanced Air Mobility

Electric Vertical Take-Off and Landing (eVTOL) vehicles represent a major advancement in aviation technology, poised to transform urban mobility and global transportation networks. This book looks at how these aircraft are being incorporated into existing and emerging aviation infrastructures worldwide. This title addresses the knowledge gap between traditional aviation practices and the innovative demands posed by eVTOL technologies and offers insights based on research and real-world applications.

This volume offers guidance for anyone involved in the design, implementation, regulation, or operation of eVTOL technologies within the evolving global aviation landscape. Exploring the subject through four distinct sections, it covers foundations and theoretical context, operations and training methodologies, economic viability and market trends, and global expansion strategies. Readers will discover specialized training approaches, operational protocols, human-machine interface considerations, regulatory frameworks, and stakeholder collaboration strategies essential for successful implementation. Featuring 20 international case studies from organizations including the FAA, EASA, IATA, ICAO, Airbus, Boeing, Volocopter, Archer, Lillium, Joby, and the Advanced Air Mobility Institute, this title delivers practical context for theoretical concepts, thus equipping readers with the knowledge to navigate the complexities of this technological advancement through a human-centric approach.

Written for a diverse audience, *Advanced Air Mobility: Human Factors, Ecosystem Readiness, and the Integration of eVTOLs* is for professionals and students in the fields of aviation management, human factors engineering, regulatory affairs, logistics, manufacturing, safety investigation, artificial intelligence, cybersecurity, and airport operations.

Advanced Air Mobility

Human Factors, Ecosystem Readiness, and the Integration of eVTOLs

Edited by
Dimitrios Ziakkas, Amin Vafadar,
and Debra Henneberry

CRC Press
Taylor & Francis Group
Boca Raton London New York

CRC Press is an imprint of the
Taylor & Francis Group, an **informa** business

Designed cover image: Shutterstock

First edition published 2026
by CRC Press
2385 NW Executive Center Drive, Suite 320, Boca Raton FL 33431

and by CRC Press
4 Park Square, Milton Park, Abingdon, Oxon, OX14 4RN

CRC Press is an imprint of Taylor & Francis Group, LLC

ISBN: 978-1-041-15562-1 (hbk)
ISBN: 978-1-041-16632-0 (pbk)
ISBN: 978-1-003-68555-5 (ebk)

DOI: 10.1201/9781003685555

Typeset in Times
by Apex CoVantage, LLC

Contents

APPENDICES

About the Editors

Dimitrios Ziakkas is Assistant Professor of Human Factors at Coventry University's School of Future Transport Engineering, UK. An Airbus captain with A-320, A-330/350, and A-340 type ratings, he brings over 35 years of military and civil aviation experience, exceeding 12,500 flight hours. After retiring from the Hellenic Air Force in 2008, he flew commercially with Olympic Air, Qatar Airways, and Legend Airlines. He holds multiple degrees, including Ph.D., M.S. Human Factors, and MBA. His research focuses on aviation technology adoption, AI-AMM training/operations, flight safety, and Advanced Air Mobility certification requirements.

Amin Vafadar is Aviation Development Specialist focusing on Advanced Air Mobility (AAM) integration in the UK. His research examines how AAM enhances regional connectivity and supports sustainable development. He holds a commercial pilot license and an MSc in Sustainable Air Transport Operations from the University of Salford, UK, and is completing a DBA in aviation development strategy. As a former leader of the Advanced Air Mobility Institute, he organized the Global AAM Forum with representatives from over 75 nations. He served as Business Development Lead at Grasshopper Air Mobility and authored "CRM Odyssey."

Debra Henneberry is Assistant Professor at Purdue University's School of Aviation and Transportation Technology, USA, with nearly 20 years of flight instruction experience. Before joining Purdue, she taught at an aeronautical college in New York City and flew as a personal pilot. She brings unique emergency response expertise, working for over 10 years as a first responder and emergency medical technician with extensive Department of Homeland Security training. Her research addresses flight training tactics, aviator talent pipeline expansion, and pilot well-being. She holds degrees from Dickinson College (USA), the University of Leicester (UK), and St. John Fisher University (USA).

Contributors

Argyrou, Aspasia
National Kapodistrian University of Athens
Greece

Bilotkach, Volodymyr
Associate Professor Purdue University
USA

Bokas, Eleftherios
Safety Manager Sky Express
Greece

Chafik, Aymane
University Polytechnica Bucuresti
Romania

Chazapis, Spiros
Human Factors Horizon
Romania

Diamantopoulos, Andreas
Sparfell
Austria

Hartley, Jam
Founder Beat Aviation
UK

Henneberry, Debra
Assistant Professor Purdue University
USA

ibne Masood, Fahad
Faculty Muscat College
Oman

Ives, Jim
Board Member AAM Institute
USA

Kosmidou, Anastasia
Human Factors Horizon
Greece

Lang, Tomislav
Founder FlyvBird (Vini)
Germany

Lekea, Ioanna
Hellenic Air Force Academy
Greece

Maleviti, Eva
Program Coordinator Embry-Riddle
USA

Natakusuma, Hans Cornelius
EVA Flight Training Academy
USA

Pechlivanis, Konstantinos
Technical University of Crete
Greece

Plioutsias, Anastasios
Coventry University
UK

Roth, Ian Ibrook Tow
Purdue University
USA

Sanchez, Gustavo A.
Bowling Green State University
USA

Scott, Benjamyn
Assistant Professor Leiden University
The Netherlands

Sloat, Daniel
Founder AAM Institute
USA

Synodinou, Dimitra
Human Factors Horizon
Greece

Tsentemeidou, Fay Roberta
Legend Airlines
Greece

Vafadar, Amin
Advanced Air Mobility Institution
Iran

Vasileiou, Vasileios
Legend Airlines
Romania

Vermaak, Gary
Senior Advisor AAM Institute
South Africa

Waseem, Khalid
Purdue University
USA

Ziakkas, Dimitrios
Assistant Professor Coventry University
UK

Preface

Every intellectual inquiry begins with a question, and this book emerged from several. How will humanity negotiate coexistence with an emergent vertical dimension of mobility? What ethical obligations accompany the technological capacity to ascend—quietly and autonomously—within urban airspace? And how can innovation in aerial mobility be governed to advance social equity, ensuring that the benefits of flight accrue to the many rather than the privileged few?

This work emerged from the belief that Advanced Air Mobility (AAM) is not solely a technological occurrence but it represents a human experiment in coexistence. Electric vertical take-off and landing (eVTOL) aircraft, digital airspace management, and autonomy are tools; their meaning emerges only when placed in the broader context of human purpose, equity, and sustainability. The chapters that follow were written not to predict the future of flight, but to question how responsibly we intend to construct it.

The research and reflections contained here are the result of interdisciplinary collaboration—bridging aeronautical engineering, systems integration, human factors, environmental ethics, and economic foresight. The writing process itself mirrored the integration challenges of AAM: technical knowledge had to converse with regulation, data had to coexist with moral reasoning, and theory had to remain answerable to experience.

I wish to express my deepest gratitude to the individuals and institutions whose trust, insight, and companionship made this endeavor possible. I am especially indebted to Dr. Debra Henneberry and PhD (C) Amin Vafadar, whose intellectual generosity and steadfast friendship exemplify the essence of scholarly mentorship. Their guidance has been less an act of instruction than a dialogue of mutual discovery—a reminder that the highest form of teaching does not merely transmit knowledge but cultivates understanding, humility, and shared purpose.

I am profoundly grateful to the Advanced Air Mobility Institute for nurturing a space where the dialogue on global mobility transcends theory to become a living experiment in collective imagination—where research, ethics, and policy converge in service of the human experience of movement. To my intellectual anchors at Coventry University (UK) and Purdue University (USA), whose partnership in aeronautical systems research revealed that innovation is not born of technology alone, but of the meeting between rigor and wonder, between inquiry and empathy. My deepest thanks also go to the National and Kapodistrian University of Athens (STIS Lab) and the Hellenic Air Force Academy, whose mentorship and moral clarity continue to affirm that the making of an aviator begins with the making of a citizen—and that flight, at its essence, is as much an ethical act as it is a technical one. Each of these institutions and their people contributed to the human-centered perspective that defines this book. Their laboratories, classrooms, and discussions shaped the conviction that technology without ethical intentionality is incomplete.

We address this volume to leaders, investors, regulators, educators, engineers, and citizens alike. It invites readers to view AAM not as a disruptive technology but as a shared civic project—a multidimensional framework of mobility, trust, and sustainability. The future of flight will not be measured solely by operational range or battery endurance, but by how inclusively and transparently we define its benefits and distribute its risks.

As we stand at the threshold of a new aerial era, the task is not to accelerate without reflection but to design progress with conscience. If these pages encourage readers to think of AAM as both an engineering achievement and a moral commitment, then the purpose of this book will have been fulfilled.

Dr. Dimitrios Ziakkas
Coventry, UK, 2025

1 Introduction to Advanced Air Mobility (AAM) and Electric Vertical Take-Off and Landing (eVTOL)

Dimitrios Ziakkas, Amin Vafadar, and Daniel Sloat

1.1 INTRODUCTION

Aviation has repeatedly reinvented how societies use the sky—first as exploration, then as global commerce, and now as an everyday mobility layer at urban-regional scales. AAM represents the latest turn in this evolution: a systems-level shift that couples aircraft innovation (electrification, distributed propulsion, and high levels of automation) with new operating concepts, infrastructure, and rulemaking. In contemporary usage, the US Federal Aviation Administration (FAA) frames AAM as the safe, efficient integration of highly automated aircraft into the National Airspace System (NAS), emphasizing that AAM is not a single technology but a collection of new and emerging technologies applied to aviation transportation (FAA, 2024a, 2024b).

1.2 CONCEPTUAL BACKGROUND AND EVOLUTION OF eVTOL AVIATION

Amin Vafadar and Daniel Sloat

The presented historical context identifies the inflection points that distinguish AAM from prior "flying car" eras and introduces the maturity frameworks and policy milestones that now signal system readiness (Figure 1.1). The aspiration to merge the door-to-door convenience of road travel with the speed of flight is as old as powered aviation. In 1917, Glenn Curtiss's Autoplane—a triplane-topped "limousine of the air"—symbolized one of the earliest attempts at a roadable aircraft. It achieved only short hops, but it captured a century-long design instinct to hybridize ground and air transport (Smithsonian Air & Space, 2018). Mid-century helicopter commuter services demonstrated a more practical urban air option. New York Airways operated scheduled services connecting midtown Manhattan with area airports until a 1977 accident on the Pan Am Building heliport—traced by the National Transportation Safety Board (NTSB) to landing-gear fatigue—ended rooftop operations and became

DOI: 10.1201/9781003685555-1

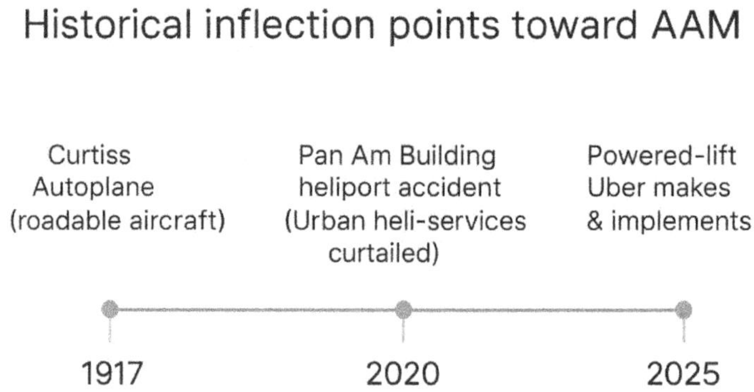

FIGURE 1.1 Historical inflection points from roadable aircraft and urban helicopters to AAM rules and readiness.

a formative reminder that Urban Air Mobility (UAM) requires reliability, rigorous maintenance, and public acceptance, not just novel aircraft (NTSB, 1977). Together, these episodes underscore two durable truths:

1. UAM is not new.
2. Scaling it demands a system (aircraft + procedures + infrastructure + social license), not just a machine.

1.2.1 The AAM Ecosystem Evolution

The AAM evolution worldwide follows a challenging engagement of multiple stakeholders that differentiate the design—planning and implementation stages. The structure of the book follows a 360-degree approach, offering a scientific-technological approach to the implementation/adaptation of AAM in society, analyzing in detail the following aspects.

- **The electrification-automation inflection:** What differentiates AAM today from previous waves is the convergence of enabling technologies and the codification of operating frameworks. Electrified propulsion and distributed-electric architectures reduce mechanical complexity and enable quieter, more failure-tolerant configurations—critical for urban acceptance. In parallel, advances in flight-control software, sensing, and human-autonomy teaming promise to reduce workload and, over time, enable new operating concepts. To organize progress, NASA's Urban Air Mobility Maturity Level (UML) scale characterizes six levels of ecosystem evolution—from early piloted demonstrations to dense, routine operations and high automation—linking operational complexity and traffic density to enabling technologies and procedures (Goodrich et al., 2021). Complementing this, the FAA's UAM Concept of Operations (ConOps) 2.0 sketches how new entrants, procedures, and infrastructure can integrate with existing air traffic services across near-, mid-, and far-term horizons (FAA, 2023a). These maturity constructs have become the lingua franca for aligning technology roadmaps with regulatory and operational readiness.

- **Policy milestones: from guidance to rules:** Policy has pivoted from exploratory guidance to binding rulemaking—a decisive step from prototype to service. In the United States, the FAA's final rule and Special Federal Aviation Regulation (SFAR) for powered-lift (published November 21, 2024; corrected January 3, 2025) establish pilot certification pathways, operating minima, and transitional frameworks that close a long-standing gap between type certification and operations for eVTOL and related aircraft (FAA, 2024c, 2025c). The rule clarifies how powered-lift operations align with existing helicopter-like and airplane-like regimes, providing the policy scaffolding for early commercial services. Globally, regulators are also moving from paper to practice: in April 2024, China's CAAC granted EHang the industry's first Production Certificate (PC) for a pilotless passenger eVTOL (EH216-S), following earlier type-certification progress—an end-to-end signal (design → production) that one jurisdiction has stood up a complete regulatory chain for this class (EHang, 2024). Europe, meanwhile, has advanced organizational approvals: Volocopter holds European Union Aviation Safety Agency (EASA) Design Organization Approval (DOA) and an extended POA explicitly covering serial production and, in 2024, received Approved Training Organisation status to ready pilots for entry-into-service (Volocopter, 2024). These milestones accelerate ecosystem investment in vertiports, training, digital services, and community engagement.
- **Demonstrated operations and ecosystem signals:** Demonstrations have moved from airshows to operationally relevant sorties in controlled airspace, including piloted eVTOL flights between public airports under FAA oversight, continued expansion of Part 135 air carrier capabilities, and integration testing of digital scheduling and flow-management tools for high-tempo on-demand operations (e.g., Joby's ElevateOS) (Joby Aviation, 2024). In parallel, multi-stakeholder programs such as the US DOT Volpe/FAA "Up, Up, and Away" thought-leadership series (2023–2024) and the World Economic Forum's AVIATE initiative (2024, 2025) have emphasized responsible deployment—safety, sustainability, and equity—alongside workforce development and interagency coordination (Volpe Center, 2024; WEF, 2024, 2025). These initiatives are less about publicity than governance capacity: they align agencies, cities, and industry around shared terminology, metrics, and phased pathways to scale.
- **From "aircraft-first" to "ecosystem-first":** AAM reframes aviation from a product to a service ecosystem with five interdependent layers:
  1. Aircraft: Electric and hybrid-electric architectures, high-assurance flight controls, and automation progressing with UML levels (Goodrich et al., 2021).
  2. Airspace & Procedures: Evolutionary integration via existing IFR/VFR constructs, strategic deconfliction, and performance-based navigation—stepping stones outlined in UAM ConOps 2.0 (FAA, 2023a).

3. *Infrastructure:* Vertiport siting, power/charging, safety zones, and community noise abatement are now prioritized by FAA's "New Entrants" infrastructure guidance (FAA, 2024b).
4. *Operations & Business Models:* Expansion of Part 135 capabilities, phased route launches, and public-private partnerships to seed corridors and demand.
5. *Governance & Public Acceptance:* Rulemaking (e.g., powered-lift SFAR), standardization, and community engagement frameworks advanced by national authorities and international coalitions (WEF, 2024, 2025).

1.2.2 Definitions and Taxonomy

Standardization in definition and taxonomy is followed in the whole book. We adopt authoritative, regulator-aligned definitions and a vehicle-system taxonomy that is consistent with current FAA, NASA, EASA, and WEF usage. All presented definitions are aligned with the regulatory bodies, highlighting any differences among them. The standardization in definitions and taxonomy follows:

A. Systems-level presentation,
B. Vehicle categories and regulatory anchors,
C. eVTOL architecture taxonomy (design families),
D. Crewing, automation, and operations modality,
E. NASA Urban Air Mobility Maturity Levels (UML).

1.2.2.1 Systems-Level Presentation

AAM denotes the transportation system that integrates highly automated, often electrically powered aircraft into civil airspace and the built environment to move people and goods. *UAM* is the intracity subset; *Regional Air Mobility (RAM)* extends to inter- and intra-regional flows, often leveraging local airports and thin-haul corridors. eVTOL architectures are grouped by the function of propulsion—vectored thrust, lift-plus-cruise, and multicopter (wingless)—and may be certificated under different categories depending on configuration and flight modes. In the United States, many eVTOLs will operate as powered-lift; in Europe, EASA's SC-VTOL provides configuration-agnostic airworthiness objectives. We reference autonomy via NASA's UML scale (1–6), which characterizes ecosystem maturity rather than only vehicle TRLs.

In this book (Table 1.1), AAM is the umbrella; UAM = urban subset; RAM = regional subset. When a source uses UAM in the broader sense, we map it explicitly to AAM to avoid ambiguity. eVTOL denotes electric aircraft with VTOL capability across the three architectures above; energy-system qualifiers (all-electric, hybrid-electric, and fuel-cell) are attached as needed. Moreover, powered-lift is a regulatory category (FAA) that may apply to many—but not all—AAM vehicles; SC-VTOL (EASA) is an airworthiness special condition applicable to VTOL-capable aircraft independent of marketing labels. Finally, references to autonomy are grounded in UML and specific regulatory guidance; we avoid marketing-level "Level N" labels except when a source defines them.

TABLE 1.1
System-level terms used throughout the book

Term	Working Definition
AAM (umbrella)	System that integrates new aircraft, services, infrastructure, and governance across urban-regional scales.
UAM (subset of AAM)	Intracity operations—initially piloted, corridor-based, and evolving toward denser networks.
RAM (subset of AAM)	Inter-/intra-regional links (≈50–300+ km) using GA airports and thin-haul corridors.

TABLE 1.2
Regulatory anchors relevant to VTOL-capable aircraft

Anchor	Scope	Implications
Powered-lift (FAA)	Heavier-than-air; vertical take-off/landing; non-rotating airfoils in horizontal flight.	Pilot certification and operations defined; intersects Parts 91/97/135/136.
SC-VTOL (EASA)	Airworthiness objectives for small VTOL-capable aircraft.	Configuration-agnostic; MoCs for propulsion, handling, and crashworthiness.
Rotorcraft (varies)	Category for some wingless multicopters/rotorcraft.	Applicability depends on design and jurisdiction.

1.2.2.2 Vehicle Categories and Regulatory Anchors

- **Powered-lift (aircraft category):** Under 14 CFR § 1.1, powered-lift means a heavier-than-air aircraft capable of vertical takeoff, vertical landing, and low-speed flight, depending principally on engine-driven lift devices or engine thrust for lift in those regimes and on non-rotating airfoils for lift in horizontal flight (FAA, 14 CFR §1.1). Recent FAA rulemaking (Part 194/SFAR and related guidance) clarifies pilot certification and operational rules for powered-lift operations (FAA, 2024b, 2024c). Note a nuance: some eVTOL designs that never become wing-borne (e.g., wingless multicopters) may not meet the §1.1 powered-lift definition and could be certificated under other categories (e.g., rotorcraft) depending on design and jurisdiction (FAA, 2024c, discussion of comments).
- **VTOL-capable aircraft (EASA "SC-VTOL"):** EASA's Special Condition for VTOL establishes airworthiness objectives for small VTOL-capable aircraft, with evolving Means of Compliance (MoC) publications—Issue 2 concluded in 2023; additional MoCs were published in 2025—covering crashworthiness, propulsion, handling qualities, and other domains. SC-VTOL is configuration-agnostic (lift+cruise, tilt-rotor/wing, multicopter), focusing on system safety and performance rather than prescribing specific architectures (EASA, 2023–2025). Table 1.2 presents the VTOL global regulatory ecosystems.

TABLE 1.3
eVTOL architecture families and primary trade-offs

Architecture	Description	Salient Trade-offs
Vectored thrust	Common propulsors tilt/re-vector between vertical and forward flight.	High cruise efficiency; mechanical complexity in tilting systems.
Lift + cruise	Dedicated lift rotors + separate cruise propulsor.	Simpler transition; cruise performance depends on wing loading and drag.
Multicopter (wingless)	Multiple fixed-axis rotors; no wings.	Excellent low-speed control; range/efficiency constrained; short hops.

Note: Power source qualifiers (e.g., all-electric, hybrid-electric, fuel-cell hybrid) describe energy systems applicable across architectures.

1.2.2.3 eVTOL Architecture Taxonomy (Design Families)

The presented taxonomy widely used in NASA/industry analyses is descriptive (engineering trade-space) rather than regulatory; designers may certify under powered-lift or rotorcraft depending on the specific configuration and flight modes (Table 1.3).

1.2.2.4 Crewing, Automation, and Operations Modality

- **Crewed versus uncrewed:** Early AAM services will be crewed/piloted (often single-pilot) under commercial operation specifications. Uncrewed operations (remote or highly automated) are expected to scale first in cargo/utility missions and later in passenger services as assurance frameworks mature (FAA, 2023b; WEF, 2024, 2025).
- **Autonomy scope:** Autonomy refers to aircraft functions (flight control, navigation, and contingency management) and system functions (fleet dispatch and strategic deconfliction). Moreover, the levels are mission and regulator-dependent; this book references autonomy *qualitatively* (human-assistive → supervised → high automation) and *quantitatively* via UML ecosystem milestones.

1.2.2.5 NASA Urban Air Mobility Maturity Levels (UML)

To align technical readiness with operational complexity and density, the research team follows NASA's UML 1–6 scale (Goodrich et al., 2021). Goodrich et al. explicitly note that UML is a projection intended to guide planning; different use cases (e.g., cargo vs. passenger and urban vs. regional) may advance along different trajectories and risk tolerances (Goodrich et al., 2021). FAA's UAM ConOps v2.0 and the AAM Implementation Plan align with this staged maturation concept (FAA, 2022, 2023c).

The presented levels are ecosystem milestones, not just vehicle TRLs. In detail:

1. *UML-1 (Initial Capability):* Limited piloted demos/prototypes; highly constrained environments; low density; day VFR; and minimal integration with ATC/UTM.

2. *UML-2 (Early Commercial Piloted Ops):* First revenue services on fixed corridors/routes; low density; refined procedures; and foundational digital scheduling/deconfliction.
3. *UML-3 (Expanded Piloted Ops):* Increased city/region coverage; moderate density; more contingency handling; and enhanced ATC/UTM interfaces.
4. *UML-4 (Assisted Autonomy at Scale):* Higher density and diversity of operations; greater automation in vehicle and traffic services; performance-based separation; and time-based flow management.
5. *UML-5 (High-Density Mixed Ops):* Widespread operations across metro regions; mixed crew states (piloted/remote/supervised autonomy); and robust resilience and off-nominal handling.
6. *UML-6 (Ubiquitous, Largely Autonomous Ops):* Routine, high-density service with extensive automation in both vehicles and traffic management; minimal human tactical intervention.

1.3 OBJECTIVES AND RATIONALE FOR GLOBAL eVTOL INTEGRATION

Amin Vafadar and Dimitrios Ziakkas

The AAM rationale is multi-dimensional (Table 1.4). Mobility efficiency arises from compressing generalized travel cost (time, reliability, and access) by opening low-altitude airspace to high-frequency passenger and cargo services. Public-interest missions—organ transport, point-of-injury response, disaster relief, and inspection—offer tangible early value and a pragmatic acceptance pathway. Environmental arguments rest on tailpipe zero emissions, credible pathways to lower life cycle carbon, and quieter soundscapes relative to conventional rotorcraft. Economically, AAM is an

TABLE 1.4
AAM objectives and the enabling policy/evidence base

Objective	What Success Looks Like	Evidence/Guidance Anchors
Mobility efficiency	Time/reliability gains on binding corridors (e.g., airport access, cross-metro links).	ConOps v2.0; Innovate28; value-of-time literature.
Public services	Medical logistics, EMS insertion/extraction, disaster mapping, and resupply.	City SUMP integration; inter-agency governance.
Environment	Local zero emissions; quieter profiles; LCA-based decarbonization trajectories.	Grid mix; duty cycles; procedure design.
Economic/industrial	Jobs, supply chains, digital services, vertiport build-out.	Workforce upskilling; PPPs; catalytic effects.
Equity & acceptance	Noise justice, coverage, and transparent safety/noise dashboards.	Community engagement: metrics shared via public portals.

innovation platform spanning manufacturing, batteries/propulsion, vertiports, digital traffic services, and maintenance ecosystems; it supports workforce development and regional competitiveness. Social legitimacy depends on safety transparency, equitable access, route siting over low-sensitivity corridors/waterways, and meaningful engagement. Finally, policy timing matters: accelerated rulemaking and guidance in the United States, Europe, and parts of Asia are moving from paper to practice, creating windows for early deployment where rules, infrastructure, and workforce mature together.

1.4 MISSION ARCHETYPES AND NEAR-TERM USE CASES

Amin Vafadar

The first wave of AAM services will not arrive as a single, monolithic product. It will take shape through a set of mission archetypes that match specific travel needs with airspace, infrastructure, and governance that are ready to support them (Table 1.5). This section presents real case studies in a continuous, human-centered account that explains what each archetype does, where it is already being demonstrated, and which constraints must be respected so that early operations earn—and keep—the public's trust.

1.4.1 Passenger Mobility

Passenger mobility is the most visible strand of the AAM tapestry and, for many readers, the most intuitive. Among passenger options, airport access shuttles are the clearest near-term candidate for routine service. The design problem is straightforward: move people reliably between a central business district or a major intermodal node and the nearest hub airport in a fraction of the surface travel time. Initial operations are piloted and deliberately conservative—fixed routes, predictable schedules, and published procedures that allow air traffic control to anticipate each movement (FAA, 2023a). Publicly announced routes such as Manhattan to Newark Liberty International Airport illustrate the model: a sub-10-minute flight tailored to commuter peaks, priced as a premium product, and supported by tightly managed ground flows at both termini (Archer Aviation & United Airlines, 2022). The technical novelty here is less important than the ecosystem fit. These services depend on certified powered-lift operations, heliport or vertiport facilities that meet contemporary siting and safety guidance, and noise management that is agreed upon with surrounding communities (FAA UAM ConOps v2.0). Without that social license—built through transparent outreach and credible monitoring—airport shuttles remain a paper exercise rather than public transport. A more ambitious but still near-term passenger application is the intra-metro air taxi.

- **Intra-metro air taxi (point-to-point):** Instead of shuttling between two known endpoints, aircraft connect a constellation of distributed pads across urban and suburban neighborhoods. In the early stages, routes remain corridor-based to simplify integration with existing air traffic flows; only as procedures, data, and public comfort mature do networks open toward more flexible patterns of service, in step with the progression from UML-2 to UML-4 (FAA, 2023a). Demonstrations in demanding terminal areas—such as procedure evaluations around Dallas-Fort Worth—have shown that well-designed tracks can coexist with Class B and C operations (FAA, 2023b). Several constraints—limitations—challenge the implementation of intra-metro air-taxi operations. Dispatch reliability is sensitive to local weather minima; turnaround times must be aligned with charging cycles; and route siting should favor river corridors, waterfronts, and industrial belts to minimize exposure of dense residential districts to repetitive overflight. When those constraints are treated as design inputs rather than afterthoughts, the air-taxi idea becomes a practical mobility layer rather than a curiosity.

- **Urban tourism/sightseeing:** A third passenger archetype—urban tourism and sightseeing—may appear modest compared with the promises of a full commuter network, but it often provides a more realistic path to public familiarity. Short circuits over iconic corridors, operated at low frequency and within tight noise envelopes, allow operators and city authorities to rehearse passenger handling, safety briefings, and ground operations in a controlled environment. The piloted flights over Singapore's Marina Bay, coordinated with a full-scale VoloPort demonstration, made this tangible by combining operational discipline with a public-facing showcase of boarding, marshaling, and turnaround practices (Volocopter & Skyports, 2019). The enablers are straightforward: scenic tracks that avoid acoustically sensitive neighborhoods, clear visual separation from busy airspace, and a visitor market willing to pay for the experience. The value lies in building trust through competence that can be observed rather than asserted. Event-bounded services operate on the same principle of rehearsal, but with an explicit deadline and a spotlight.
- **Expo/event demonstration services:** Time-limited routes during major expos and international gatherings validate the interfaces between aircraft, vertiport layouts, and passenger communications under the scrutiny that such venues attract. Plans for Osaka Expo 2025, including public flights and additional demonstrations at OSAKAKO Vertiport, capture how these projects both test procedures and educate the public (SkyDrive, 2025). Because the window is short and the operating context is unique, these services rely on targeted regulatory permissions, precisely timed NOTAMs, and dedicated emergency-response coverage. They will never move many people per hour, but they can move opinion by showing what "good" looks like in practice.

1.4.2 Regional Connectivity (RAM)

- **Short-haul intercity shuttles (50–200+ km):** Beyond city centers, regional connectivity—often described as RAM—extends the logic of AAM to thin-haul corridors of 50 to more than 200 kilometers. The concept is to pair secondary cities or town-to-town links using underutilized general aviation airports. Depending on stage length and payload, these flows may use eVTOL aircraft or electric short-takeoff-and-landing designs (Borer et al., 2023). State and regional planning documents, such as Washington State's 2025 aircraft plan, already call out passenger and cargo shuttles that complement highways and conventional feeder services (WSDOT, 2025). The gating factors are unsurprising: adequate energy density or hybridization for range and weather robustness for dispatch. Early business cases tend to blend commuter seats with high-value light cargo to stabilize utilization and cash flow until economies of scale materialize.

1.4.3 Cargo and Logistics

- **Small-parcel last-mile delivery (sUAS):** Cargo and logistics provide the most mature AAM operations today and will remain essential to the sector's credibility. At the small-parcel end, on-demand delivery from retail nodes to households—often moving one to five kilograms—has proven scalable in geographies where regulators have authorized routine operations under Part 135 or equivalent. UPS Flight Forward's standard certificate and thousands of medical-campus flights since 2019 are well-documented (FAA, 2019). In parallel, partnerships such as Wing and Walmart's expansion across the Dallas-Fort Worth area signal how retail logistics can reach large household catchments when the legal basis is clear (Walmart, 2024; Wing, 2023, 2025). Continued progress hinges on a stable, scalable framework for beyond visual line-of-sight operations; the FAA's 2025 NPRM sketches that pathway, and with it the unit-economics improvements that come from higher autonomy (FAA, 2025a).
- **Middle-mile/urgent logistics (larger UAS/eVTOL):** Middle-mile/urgent logistics occupy a different niche. Here, larger UAS or eVTOL aircraft shuttle between hubs or specialized facilities—hospitals, laboratories, offshore platforms—where the value of speed and predictability justifies the service. The Matternet M2's type and production certificates (both achieved in 2022, with subsequent amendments) have enabled scaled medical logistics and illustrate what a certified pipeline looks like for this class (Matternet, 2022). Success depends on site-specific risk mitigations, secure chain-of-custody, and communications resilient to urban multipath and rural coverage gaps through layered cellular/5G and GNSS augmentation.

1.4.4 Public Service and Emergency Response

- **Medical logistics and time-critical health services:** Public service and emergency response missions articulate the strongest societal rationale for AAM and often furnish the most persuasive acceptance pathways. Medical logistics—blood, vaccines, pathology samples, and critical devices—already show how protected aerial corridors can compress clinically significant timelines, especially when surface networks are congested. Zipline's work in Ghana during the COVID-19 pandemic, which cut sample transport times from hours to under one hour, made the case that reliability is not a slogan but the product of disciplined operations (Lamptey, 2020). Scaling this kind of mission requires careful hospital integration, documented chain-of-custody, and weather contingencies, all supported by simple, standardized pads and automated release and recovery systems that reduce ground-handling variability.

- **Medical logistics and time-critical health services:** At the more acute end of the spectrum, air-ambulance and point-of-injury response can complement helicopters with quieter eVTOL profiles in congested corridors. The AiRMOUR program offers a governance and operational playbook for integrating medical UAM into city planning instruments such as Sustainable Urban Mobility Plans (AiRMOUR, 2023a), while NASA's public-services white paper enumerates the aircraft and mission implications in the language of requirements rather than aspirations (Doo et al., 2021).
- **Disaster response, firefighting, and search and rescue:** Disaster response, firefighting, and search and rescue add further layers: uncrewed aircraft for intelligence, surveillance, and reconnaissance; mapping and supply drops; and, where justified, piloted eVTOL insertions for personnel movement in austere conditions. The FAA's infrastructure planning guidance recognizes these roles, and NASA has highlighted the cost and speed advantages for public missions (Doo et al., 2021; FAA, 2024a). Even so, constraints remain; specialized training, payload and temperature limits, and standards for aircraft rescue and firefighting at vertiports are still maturing.
- **Early autonomous passenger services (restricted contexts):** As far as the pace of autonomy will not be uniform across all use cases, some jurisdictions have chosen to authorize early autonomous passenger services in narrowly defined contexts. These are typically short, geofenced sightseeing routes under close local oversight, supported by centralized fleet supervision and carefully managed airspace. The EH216-S experience in China—where type, production, and operating certificates have enabled touristic flights in selected cities and a handful of overseas demonstrations—illustrates the potential. It also underscores the caveat: what is permissible in one regulatory environment is not automatically portable to another, especially where integration with instrument flight rules and higher-density traffic is concerned. Certification scope, operating envelope, and contingency handling are ultimately jurisdiction-specific; generalizing from showcase routes to mainstream public transport is neither automatic nor quick.
- **Infrastructure and utility services:** A final family of missions—inspection and utility services—tends to attract less attention but delivers high societal return (Table 1.6). Routine inspection of powerlines, pipelines, railways, bridges, ports, and offshore assets benefits from the repeatability, safety, and speed that purpose-built aircraft can deliver. These operations are already embedded in the FAA's UAM ConOps 2.0 and routinely appear in state-level

AAM plans; their growth tracks the maturation of beyond visual line of sight (BVLOS) rules and the emergence of standardized data pipelines that let inspection data flow straight into asset-management systems (FAA, 2023a). Because these missions are less sensitive to passenger perceptions, they can accumulate operational hours that, in turn, provide the safety evidence and maintenance learning curves the broader ecosystem needs.

If one pulls the thread through these archetypes, three cross-cutting realities emerge:

Firstly, the earliest passenger services likely to deploy at scale are those that remain on fixed routes—airport shuttles and limited tourism—where vertiports, ATC procedures, and community expectations can be tightly controlled (FAA, 2023b).

Secondly, cargo and medical logistics are not hypothetical; they are operating now under Part 135 and other authorities, and they will expand as BVLOS pathways stabilize (FAA, 2019, 2025b).

Thirdly, public-good missions strengthen legitimacy when they are woven into existing emergency frameworks rather than treated as stand-alone pilots (AiRMOUR, 2023a).

Table 1.5 summarizes the Key Performance Indicators (KPIs) and enabling conditions for early mission archetypes. The KPIs are presented by archetype—stage length, payload, block time, sortie rate, dispatch, noise objective, regulatory basis, and infrastructure dependencies.

The following Comparative Readiness Matrix (2025–2030) presents the technology readiness of the global ecosystem.

1.5 ECOSYSTEM OVERVIEW

Amin Vafadar

AAM should be understood as a system-of-systems rather than a discrete product. Its promise rests on the disciplined coupling of aircraft technology with procedures and institutions: vehicles are certified and supported; airspace services evolve to accommodate new operations; vertiports and power systems are planned as energy assets; operators are authorized and audited; digital networks carry data securely and interoperably; and governance frameworks protect safety, equity, and environmental performance as services scale. This overview traces each layer and highlights the current anchors—standards, advisory circulars, concepts of operations, and guidance—that convert promising prototypes into reliable transport (EASA, 2024; FAA, 2023c; WEF, 2024, 2025).

- **Vehicles and certification pathways:** At the vehicle level, several certification pathways now exist, reflecting the diversity of configurations that fall under the AAM umbrella. In the United States, many electric vertical

TABLE 1.5
Presentations of KPIs and enabling conditions for early mission archetypes

Archetype	Stage Length (km)	Payload	Block Time (min)	Sorties/hr (per a/c)	Dispatch Target (%)	Noise Objective	Regulatory Basis	Infra dependencies
Airport access shuttle	10–30	2–6 pax	8–12	2–4	£95	<65 dBA @ 100 m	Part 135 + powered-lift ops	Vertiport Class A; 300–1,000 kW DC/stand
Intra-metro air taxi	10–40	2–6 pax	10–16	2–3	£92	Urban profile; route over water/industrial	Part 135 (piloted) + corridor procedures	Distributed pads; turnaround *10–15 min
Urban tourism/sightseeing	8–20	2–4 pax	8–15	3–5	£97	60–65 dBA @ 100 m; tight hours	Local VFR approvals; heliport standards	Limited pads; visitor processing
Event demonstration service	5–20	2–4 pax	6–12	2–4	£95	Special procedures; evert windows	Targeted waivers; NOTAM windows	Dedicated EMS cover; temporary pads
Intercity RAM shuttle	50–200+	2–9 pax	20 60+	1–2	£90	Airport-compatible profiles	Part 135/commuter; eVTOL or eSTOL	GA airports; charging or hybrid
Small-parcel last-mile (sUAS)	2–15	1–5 kg	5–12	6–10	£97	Neighborhood limits; s60–65 dBA	Part 135 (cargo) + BVLOS pathway	Roof/lot nodes; lockers; FRIA/RID
Middle-mile/urgent logistics	10–80	10 100+ kg	8–40	2–4	£95	Industrial corridors; hours managed	Part 135 (cargo) + BVLOS pathway	Warehouse hubs; secure hand-off
Medical logistics	5–50	2–20 kg	5–25	3–6	£99	Hospital corridors; priority routing	Part 135 (cargo) + BVLOS pathway	Hospital pads; chain-of-custody
Inspection & utility services	5–100+ (corridor)	Sensors/small payloads	10–60+	2–6	£95	Route over ROWs; off-peak hours	BVLOS pathway; UTM/U-space services	Access pads; data pipelines

TABLE 1.6
AAM mission families and near-term enablers

AAM Mission Families and Near-Term Enablers (2025–2030 Readiness)

Piloting Status	Corridor Control	Infrastructure Maturity
✓ Passenger mobility ✓ Regional connectivity (RAM) ✓ Cargo & logistics ✓ Public service & emergency response ✓ Early autonomous passenger services ✓ Infrastructure & utility services	✓ Passenger mobility ✓ Regional connectivity (RAM) ✓ Cargo & logistics ✓ Public service & emergency response ✓ Early autonomous passenger services ✓ Infrastructure & utility services	✓ Passenger mobility ✓ Regional connectivity (RAM) ✓ Cargo & logistics ✓ Public service & emergency response ✓ Early autonomous passenger services ✓ Infrastructure & utility services
BVLOS /Legal Status	**Noise Profile**	**Digital Operations**
✓ Passenger mobility ✓ Regional connectivity (RAM) ✓ Cargo & logistics ✓ Public service & emergency response ✓ Early autonomous passenger services ✓ Infrastructure & utility services	✓ Passenger mobility ✓ Regional connectivity (RAM) ✓ Cargo & logistics ✓ Public service & emergency response ✓ Early autonomous passenger services ✓ Infrastructure & utility services	✓ Passenger mobility ✓ Regional connectivity (RAM) ✓ Cargo & logistics ✓ Public service & emergency response ✓ Early autonomous passenger services ✓ Infrastructure & utility services

Note: Ticks indicate credible readiness in constrained contexts by 2025–2030; specificity varies by jurisdiction and mission profile.

takeoff and landing designs will be certificated and operated as powered-lift aircraft, a regulatory container that clarifies how such vehicles fit within existing operational parts and pilot qualification frameworks. The FAA's final rule and accompanying SFAR establish pilot certification and operational provisions for powered-lift—explicitly mapping how operations intersect with Parts 91, 97, 135, and 136 and creating a 10-year window to stand up pilots and operators (FAA, 2024e, 2025c). Advisory circulars translate these rules into practice: AC 21.17-4 provides type-certification guidance for powered-lift, while AC 194-1 and AC 194-2 describe operational and training templates that manufacturers and operators can implement without guesswork (FAA, 2024e). In Europe, EASA's SC-VTOL defines configuration-agnostic airworthiness objectives for small VTOL-capable aircraft, with MoC that continue to mature as evidence accumulates. Throughout this book, we use eVTOL to describe the technology, powered-lift to denote the U.S. regulatory category many (but not all) eVTOLs will use, and SC-VTOL to refer to EASA's airworthiness framework. Harmonization across authorities is advancing but not complete; the state of that convergence will shape cross-border operations and training equivalencies (EASA, 2024; FAA, 2024d).

- **Airspace services and traffic management:** Operational integration proceeds evolutionarily. In the near term, piloted AAM flights enter today's air traffic system using familiar VFR/IFR procedures, with dedicated routes or corridors where these improve predictability, and performance-based separation as densities increase—an approach codified in the FAA's UAM ConOps v2.0 (FAA, 2023c). For uncrewed and highly automated operations at scale, UTM/U-space services provide the digital substrate for authorization, strategic deconfliction, conformance monitoring, and information exchange among service suppliers. Standards such as ASTM F3548-21 (UAS Service Supplier interoperability) and ASTM F3411-22a (Remote ID performance) define the minimum connective tissue, while deployments like the InterUSS Discovery and Synchronisation Service demonstrate how federated systems can work in shared airspace (ASTM, 2021, 2022a; InterUSS, 2024). The practical implication is straightforward: early piloted passenger services will fly under conventional ATM with tailored procedures; cargo/utility fleets and later high-automation passenger services will lean more heavily on UTM/U-space constructs as enabling rules and infrastructures mature (EU 2021/664; FAA, 2023c).
- **Ground infrastructure and energy systems:** AAM's ground layer is at once airfield and power plant. The FAA's Engineering Brief 105 (2022) provided the first federal guidance for vertiport design. With EB 105A (December 2024), vertiports are reframed explicitly as a subclass of heliports, accompanied by supplemental guidance to AC 150/5390-2D (2023b) on classification, touchdown and lift-off areas, markings, parking, and downwash/outwash protections (FAA, 2022, 2023c). In parallel, ASTM F3423/F3423M-22 offers a standard specification that municipalities and project

sponsors can adopt or adapt, accelerating local planning (ASTM, 2022b). On the energy side, the National Renewable Energy Laboratory's 2023 study quantified the implications of eVTOL charging for distribution networks: across surveyed designs, DC peak demand commonly ranges from ~300 kW to ~1 MW per stand, with typical aircraft footprints on the order of 15 m × 15 m (NREL, 2023a). Multi-stand sites therefore require transformer sizing, feeder upgrades, and demand-charge management strategies (often via managed charging, storage, or microgrids). Competent infrastructure planning treats vertiports as energy assets from day one—integrating power studies, acoustic footprints, ARFF provision, security, multimodal access, and community zoning into a single siting conversation (ASTM, 2022b).

- **Operators, approvals, and business processes:** Where aircraft and infrastructure meet the travelling public, operator certification and day-to-day processes determine whether AAM feels like transport or a trial. In the United States, near-term passenger services are expected to operate under Part 135 or analogous frameworks elsewhere. The powered-lift rule clarifies how those operations intersect with Parts 91/97/135/136 and establishes a training and qualification scaffold for pilots and instructors (FAA, 2024c, 2025c). At the ecosystem level, the Innovate28 portfolio coordinates inter-agency and industry work so that one or more sites can demonstrate at-scale operations by 2028. Regardless of jurisdiction, credible operators will stand up robust safety management systems, continuing airworthiness programs, and digital flight-operations platforms for scheduling, dispatch, and performance monitoring—aligned to ConOps and the relevant advisory circulars.
- **Digital backbone, data, and cybersecurity:** Even when aircraft are piloted, AAM behaves like a digital network. Fleet management (scheduling, dispatch, and health monitoring), airspace connectivity (UTM/ATM data exchange, DSS discovery, NOTAMs), vertiport energy management, and public dashboards for noise and environmental performance all depend on well-specified data exchanges and strong security. Here again, standards do the heavy lifting: ASTM F3548 and F3411 define interoperability and identification layers, and the FAA's Remote ID rule (Part 89) is fully in effect, with toolkits for industry and FRIA carve-outs (ASTM, 2021, 2022a). Success at scale turns on API-level interoperability, secure identity, and auditability across operators, vertiports, and authorities—not on bespoke integrations that collapse under operational stress.
- **Governance: safety, sustainability, equity, and public acceptance:** AAM will scale only where it earns public legitimacy. The World Economic Forum's AVIATE work distills a durable consensus: responsible deployment rests on three pillars—safety and security; societal and environmental sustainability; and public trust and equity—supported by shared metrics and transparent dashboards (WEF, 2024, 2025). Independent studies indicate initially positive attitudes toward UAM when safety, noise, and route siting are credibly addressed (Airbus UTM, 2019; EASA, 2021). In practical terms, that implies acceptance-by-design: corridors routed over

low-sensitivity areas or waterways; clear, evidence-based noise modeling; equitable access in service planning; and early, ongoing community engagement. The international picture is consistent. In Europe, U-space rules (EU, 2021/664) are already operational, defining services and roles for digital airspace management; in the United States, UAM ConOps v2.0 lays out the evolutionary path for integration. Both anticipate phased increases in density as digital services and procedures mature (EU, 2021; FAA, 2023a).
- **Insurance, finance, and risk:** As activity moves from pilots to networks, the insurability of AAM hinges less on promises and more on data. Underwriting will increasingly rely on credible operational histories, sound cyber posture, supply-chain integrity, and transparent exposure to noise and environmental externalities. Insurance leaders already situate AAM within broader mobility risk trends and call for data-driven frameworks that reward operators who can surface standardized safety metrics, predictive-maintenance evidence, and security audit trails (Munich Re/ERGO, 2025). In short, the cost of capital and coverage will follow the quality of evidence.
- **Workforce, training, and human performance:** People remain at the center of the system. The powered-lift rule and associated advisory material establish a structured pipeline for pilots and instructors, while maintenance and ground-operations staff will require new curricula for high-voltage systems, distributed propulsion, and software assurance. Human performance must be designed for high-tempo, low-altitude operations with novel failure modes: cockpit interfaces, procedures, and training regimes should distribute workload sensibly, support resilient recovery from off-nominal events, and sustain safety culture at scale (FAA, 2023b).

In summary, AAM succeeds where the pieces are co-designed. Vehicles and procedures align with UAM ConOps; vertiports satisfy EB 105A/AC 150/5390-2D and ASTM F3423; digital systems implement ASTM F3548/F3411 and Part 89; operators are certified under the powered-lift rule and manage risk through robust SMS; and governance centers safety, sustainability, and equity with public-facing metrics (WEF, 2024, 2025). When these layers mature together, the result is not a collection of demonstrations but a dependable public service worthy of the communities it intends to serve.

1.6 SCOPE AND AUDIENCE OF THE BOOK

Dimitrios Ziakkas, Debra Henneberry, and Amin Vafadar

This book is written as both a graduate-level text and a practitioner's reference. Its purpose is straightforward: to bring the authoritative record of rules and guidance into conversation with the best available research and the lived realities of implementation. Throughout, we work from primary sources—FAA, EASA, International Civil Aviation Organization (ICAO), and ASTM materials; agency studies from NASA and NREL; and peer-reviewed or otherwise high-quality industry and academic analyses—and translate them into the decisions that regulators, cities, operators, and investors must make. The intended readership is correspondingly broad. Civil aviation authorities and air navigation service providers will find clear mappings from statutory intent to operational practice. City and infrastructure planners will see how vertiport siting, energy provision, and acoustic footprints translate into zoning, permitting, and project delivery. Operators and airport authorities will be able to trace certification pathways, operating specifications, and safety management obligations into executable procedures. Original equipment manufacturers and their suppliers will find a stable vocabulary for linking design choices to certification and integration constraints. Faculty, students, and professional learners gain a systems view that sits comfortably in a graduate seminar but remains usable in the field. Finance and insurance professionals are given the risk signals—safety data, cyber posture, supply-chain integrity, and environmental exposure—that matter for underwriting and capital deployment.

The book assumes a working familiarity with basic aeronautics, the structure and language of air traffic management, and the elements of transport economics. Where specialized terms are unavoidable, they are defined at first use and collected in a glossary aligned with official definitions so that readers can move across chapters without losing the thread. Our aim is not to sell a vision but to equip the reader—whatever their role—with a credible, evidence-based way to act.

1.7 STRUCTURE AND ORGANIZATION OF THE BOOK

Dimitrios Ziakkas, Debra Henneberry, and Amin Vafadar

The architecture of the book follows a systems-first logic. We organize material so that technology, procedures, infrastructure, and governance mature together, mirroring the progression captured by NASA's UML and the evolutionary integration path outlined in the FAA's UAM Concept of Operations and Innovate28 timeline. Each chapter opens with a compact abstract and a small set of learning objectives. The body of the chapter then weaves core narrative with "standards" boxes that pin key arguments to specific rules, advisory circulars, or MoC; "methods" boxes that set out models, checklists, or decision procedures; brief case vignettes that show what good practice looks like in context; and evidence tables that consolidate parameters and sources. Chapters close with key terms and a curated reference list so that readers can verify claims and extend their study.

Chapter 1 establishes concepts, definitions (AAM/UAM/RAM), the rationale for near-term deployment, foundational use-case archetypes, and the ecosystem lens used throughout.

Chapter 2 assembles the global regulatory picture—U.S. powered-lift rules and SFAR provisions, EASA's SC-VTOL with associated MoC, pilot licensing and operator approvals, vertiport guidance, and the roles of U-space/UTM—while noting where harmonization is advancing and where it still needs work.

Chapter 3 turns to operational integration in the air and the city: VFR/IFR procedures and corridor design, separation, and flow concepts, UTM/U-space services and contingency handling, with a careful treatment of route siting over water and low-sensitivity corridors.

Chapter 4 focuses on people—human-machine integration, competency-based training and evidence-based training for powered-lift, workload and automation management, maintenance and ground-operations upskilling, and the practices that sustain safety culture.

Chapter 5 develops the economic spine: cost structure and utilization, revenue archetypes, dispatch reliability targets, and financing vehicles from PPPs to green instruments, all tied to sandbox-to-scale roadmaps.

Chapter 6 surveys the enabling technologies—propulsion and energy systems, flight-control and autonomy assurance, the communications-navigation-surveillance stack, and digital operations platforms—emphasizing how readiness constraints shape procedures and infrastructure.

Chapter 7 addresses the institutional fabric: the roles of authorities, ANSPs, cities, operators, OEMs, and standards bodies; the consortia and memoranda that make collaboration real; and the data-sharing models required for scale.

Chapter 8 treats scaling as an engineering and civic problem—moving from minimum-viable networks to dense meshes; site selection, capacity planning, and charging scale-up; intermodal integration; and the exit criteria and KPIs that mark genuine readiness.

Chapter 9 gathers sustainability, cybersecurity, and ethics into a single governance frame—life cycle assessment and CO_2e methods, noise mapping and abatement, zero-trust patterns for cyber-risk, and equity-by-design with community acceptance practices.

Chapter 10 looks forward, developing scenarios to 2035 and 2050, tracing autonomy trajectories, examining macroeconomics and externalities, and setting out policy options and the societal commitments implied by UML-5/6 operations.

Chapter 11 offers comparative case studies across regions, extracting lessons on policy, infrastructure, acceptance, and economics that travel well.

Chapter 12 closes with synthesis: stepwise roadmaps for regulators, cities, operators, and investors, and a research agenda with metrics for accountability.

Different readers will prefer different paths. A regulator or ANSP might read from the opening chapter through the regulatory foundations and operational integration, then move directly to collaboration, scaling, and the concluding roadmaps, and consulting the sustainability chapter for governance audits. A city or airport team may step from concepts to operational integration and economics, then to scaling and governance, before turning to the synthesis. Operators and investors tend to start with the business spine, then reach back into technology and forward into collaboration and scaling, before reading the outlook and case studies with an eye to transferability. Academic readers can lift a coherent course sequence—definitions and maturity framing; regulatory and operational foundations; enabling technologies; ethics and sustainability; cases—supplemented by problem sets that mirror real tasks: vertiport appraisals that integrate geometry, power, and acoustics; corridor concepts of operations and public briefings; and business-case sensitivities that make explicit the link between utilization, dispatch reliability, and cost of capital.

A proposed way of reading is as follows:

- Policy/regulator/ANSP path: Chapters 1 → 2 → 3 → 7 → 8 → 12 (with 9 for governance audits).

- City/airport/infra path: Chapters 1 → 3 → 5 → 8 → 9 → 12.
- Operator/investor path: Chapters 1 → 5 → 6 → 7 → 8 → 10 → 11.
- Academic/teaching path: Chapters 1 (definitions/maturity) → 2 and 3 (frameworks) → 6 (technology) → 9 (ethics/sustainability) → 11 (cases), with problem sets from 5 and 8.

Finally, although this is a print-first book, it invites *active use*. Decision trees clarify certification and operating-specification pathways; process swimlanes make dispatch and flow management visible; maps and noise-contour figures anchor community conversations; and rule-to-AC-to-MoC crosswalks serve as checklists in the field. Where appropriate, QR codes point to living glossaries, updated ConOps diagrams, and open datasets so that readers can follow the evidence as standards evolve without losing the stability that a peer-reviewed text is meant to provide.

REFERENCES

AiRMOUR. (2023a). *Guidebook for urban air mobility integration: Process & management (Deliverable 6.4).* European Union Horizon 2020. https://airmour.eu/guidebook-for-uam-integration-deliverable-6-4/

AiRMOUR. (2023b). *Integration of UAM services in sustainable mobility and transport planning (Deliverable 4.5).* European Union Horizon 2020. https://www.easa.europa.eu/en/downloads/139800/en

Archer Aviation. (2025, April 17). *Vision for New York air taxi network, in partnership with United Airlines.* https://investors.archer.com/news/news-details/2025/Archer-Unveils-Vision-for-New-York-Air-Taxi-Network-Including-Routes-Between-Manhattan-and-Nearby-Airports-in-Partnership-with-United-Airlines/default.aspx

Archer Aviation & United Airlines. (2022, November 10). *First commercial electric air taxi route in the U.S.: Downtown Manhattan to newark liberty international airport.* https://investors.archer.com/news/news-details/2022/Archer-and-United-Airlines-Announce-First-Commercial-Electric-Air-Taxi-Route-in-the-US-Downtown-Manhattan-to-Newark-Liberty-International-Airport/default.aspx

ASTM International. (2021). *F3548-21: Standard specification for UAS Traffic Management (UTM) UAS Service Supplier (USS) interoperability.* https://www.astm.org/f3548-21.html

ASTM International. (2022a). *F3411-22a: Standard specification for remote ID and tracking.* https://www.astm.org/f3411-22a.html

ASTM International. (2022b). *F3423/F3423M-22: Standard specification for vertiport design.* https://www.astm.org/f3423_f3423m-22.html

Borer, N. K., Blaesser, N. J., & Patterson, M. D. (2023). *Regulatory considerations for future regional air mobility aircraft (NASA/TM).* NASA. https://sacd.larc.nasa.gov/sacd/wp-content/uploads/sites/167/2021/04/2021-04-20-RAM.pdf

Bureau d'Enquêtes et d'Analyses pour la sécurité de l'aviation civile. (2013). *Report—Serious incident: Approach above glide path, interception of ILS sidelobe signal, increase in pitch angle commanded by autopilot.* https://skybrary.aero/sites/default/files/bookshelf/2466.pdf

Doo, J. T., et al. (2021). *Electric vertical takeoff and landing (eVTOL) aircraft technology for public services: A white paper.* NASA/VFS. https://ntrs.nasa.gov/api/citations/20205000636/downloads/2021-08-20-eVTOL-White-Paper-Final_V48.pdf

EHang. (2024, April 7). *EHang secures production certificate from CAAC, clearing path for mass production of EH216-S pilotless eVTOL aircraft.* https://ir.ehang.com/news-releases/news-release-details/ehang-secures-production-certificate-caac-clearing-path-mass

European Union. (2021). *Commission implementing regulation (EU) 2021/664 on a regulatory framework for U-space*. https://eur-lex.europa.eu/legal-content/EN/TXT/PDF/?uri=CELEX%3A32021R0664

European Union Aviation Safety Agency (EASA). (2021, May). *Study on the societal acceptance of urban air mobility in Europe*. https://www.easa.europa.eu/sites/default/files/dfu/uam-short-report.pdf

European Union Aviation Safety Agency (EASA). (2024, May 29). *Easy access rules for U-space (Regulation (EU) 2021/664)*. https://www.easa.europa.eu/en/document-library/easy-access-rules/easy-access-rules-u-space-regulation-eu-2021664

European Union Aviation Safety Agency (EASA). (2023–2025). *Special condition for VTOL; means of compliance (Issue 2 and subsequent publications)*. https://www.easa.europa.eu/en/downloads/142240/en

Federal Aviation Administration (FAA). (2019, October 1). *FAA certification—UPS flight forward obtains part 135 standard*. https://www.faa.gov/newsroom/us-transportation-secretary-elaine-l-chao-announces-faa-certification-ups-flight-forward

Federal Aviation Administration (FAA). (2022, September 21). *Engineering brief 105: Vertiport design*. Washington, DC: FAA Airports. https://nbaa.org/wp-content/uploads/aircraft-operations/emerging-technologies/advanced-air-mobility-aam/20220921-FAA-Engineering-Brief-105-Vertiport-Design.pdf

Federal Aviation Administration (FAA). (2023a, January 5). *AC 150/5390-2D: Heliport design*. Washington, DC: FAA Airports. https://www.faa.gov/airports/resources/advisory_circulars/index.cfm/go/document.current/documentnumber/150_5390-2

Federal Aviation Administration (FAA). (2023b, April 26). *Urban Air Mobility concept of operations v2.0*. Washington, DC: FAA NextGen. https://www.faa.gov/sites/faa.gov/files/Urban-Air-Mobility-Concept-of-Operations-2.0.pdf

Federal Aviation Administration (FAA). (2023c, July 18). *Advanced Air Mobility implementation plan (Innovate28)*. Washington, DC: FAA. https://www.faa.gov/air-taxis/implementation-plan

Federal Aviation Administration (FAA). (2024a). *AIM, Ch. 11 §6: Advanced Air Mobility (AAM)*. https://www.faa.gov/air_traffic/publications/atpubs/aim_html/chap11_section_6.html

Federal Aviation Administration (FAA). (2024b). *Advanced Air Mobility infrastructure (Airports—New Entrants)*. https://www.faa.gov/airports/new_entrants/aam_infrastructure

Federal Aviation Administration (FAA). (2024c, November 21). *Integration of powered-lift: Pilot certification and operations; miscellaneous amendments related to rotorcraft and airplanes (Final rule)*. Federal Register. https://www.federalregister.gov/documents/2024/11/21/2024-24886/integration-of-powered-lift-pilot-certification-and-operations-miscellaneous-amendments-related-to

Federal Aviation Administration (FAA). (2024d, November 21). *AC 194-2: Pilot training and certification for powered-lift operations*. Washington, DC: FAA AFS. https://www.faa.gov/regulations_policies/advisory_circulars/index.cfm/go/document.information/documentID/1042955

Federal Aviation Administration (FAA). (2024e, December 27). *Engineering brief 105A: Vertiport design—Supplemental guidance to AC 150/5390-2D*. Washington, DC: FAA Airports. https://www.faa.gov/airports/engineering/engineering_briefs/eb_105a_vertiports

Federal Aviation Administration (FAA). (2025a). *Beyond visual line of sight (BVLOS) proposed rule (NPRM)*. https://www.faa.gov/newsroom/BVLOS_NPRM_website_version.pdf

Federal Aviation Administration (FAA). (2025b, January 3). *Integration of powered-lift: Pilot certification and operations (Correction to final rule)*. Federal Register. https://www.federalregister.gov/documents/2025/01/03/2024-30331/integration-of-powered-lift-pilot-certification-and-operations-miscellaneous-amendments-related-to

Federal Aviation Administration (FAA). (2025c, March 19). *Remote identification of drones—Part 89 resources*. https://www.faa.gov/uas/getting_started/remote_id

Goodrich, K. H., Olson, C. A., Homola, J., Rios, J. L., & Antcliff, K. R. (2021). *Description of the NASA Urban Air Mobility Maturity Level (UML) scale (NASA/TP-20205010189)*. NASA. https://ntrs.nasa.gov/api/citations/20205010189/downloads/UML%20Paper%20SciTech%202021.pdf

InterUSS Platform. (2024). *UTM/U-space deployments and DSS interoperability*. https://interussplatform.org/utm-u-space-deployments/

Joby Aviation. (2024, June 20). *FAA authorizes use of ElevateOS for air-taxi operations*. Reuters. https://www.reuters.com/business/aerospace-defense/joby-aviation-says-faa-gives-nod-in-house-software-air-taxi-operations-2024-06-20/

Lamptey, E. (2020). *The use of Zipline drones technology for COVID-19 samples transportation in Ghana. HighTech and Innovation Journal* (Special Issue). https://www.exemplars.health/-/media/files/egh/resources/ecr/ghana/75-the-use-of-zipline-drones-technology-for-covid19-samples-transportation-in-ghana-1.pdf

Matternet. (2022, November 30). *FAA production certificate for M2 drone; first U.S. UAS type certification (September 2022)*. https://www.matternet.com/newsroom/faa-production-certificate-zgbkj-dkzdc

Matternet. (2023, November 2). *M2 type certificate amendment*. https://www.businesswire.com/news/home/20231102809938/en/Matternet-Receives-First-FAA-Type-Certification-Amendment-for-its-M2-Drone-Delivery-System

Munich Re & ERGO. (2025). *Tech trend radar 2025: The future of insurance*. https://www.munichre.com/en/company/innovation/tech-trend-radar-2025.html

National Renewable Energy Laboratory (NREL). (2023a). *Life-cycle assessment case study for eVTOL services*. https://www.nrel.gov/news/detail/program/2023/nrel-and-joby-aviation-partnership-spotlights-green-ride-hailing-flight-services

National Renewable Energy Laboratory (NREL). (2023b). *Vertiport electrical infrastructure and planning considerations (NREL/TP-86245)*. https://docs.nrel.gov/docs/fy24osti/86245.pdf

National Transportation Safety Board (NTSB). (1977). *Aircraft accident report: New York Airways, Sikorsky S-61L, Pan Am Building Heliport, New York, May 16, 1977 (NTSB-AAR-77-9)*. https://www.ntsb.gov/investigations/AccidentReports/Reports/AAR7709.pdf

National Transportation Safety Board (NTSB). (2014a). *Asiana airlines flight 214 accident report*. [Report].

National Transportation Safety Board (NTSB). (2014b). *Crash during a nighttime nonprecision instrument approach to landing, UPS Flight 1354, Airbus A300-600, N155UP, Birmingham, Alabama, August 14, 2013*. https://skybrary.aero/sites/default/files/bookshelf/2945.pdf

Smithsonian Air & Space. (2018, September). *So that's what flying cars are for (Curtiss Autoplane context)*. https://www.smithsonianmag.com/air-space-magazine/13_sep2018-so-s-what-flying-cars-are-180969917/

Volocopter. (2024, February 29). *Volocopter receives POA extension for VoloCity serial production*. https://www.volocopter.com/en/newsroom/vc-poa-extension

Volocopter & Skyports. (2019, October 22). *Piloted air taxi flight over Singapore's Marina Bay & full-scale VoloPort demonstration*. https://verticalmag.com/press-releases/volocopter-air-taxi-flies-over-singapores-marina-bay/

Volpe National Transportation Systems Center. (2024, February 15). *Up, up, and away: Innovations in Advanced Air Mobility—Series finale (DOT/FAA thought-leadership series)*. https://rosap.ntl.bts.gov/view/dot/77200

Walmart. (2024, January 9). *Largest drone-delivery expansion of any U.S. retailer—1.8 million DFW households*. https://corporate.walmart.com/news/2024/01/09/sky-high-ambitions-walmart-to-make-largest-drone-delivery-expansion-of-any-us-retailer

Washington State Department of Transportation (WSDOT). (2025, June). *Advanced Air Mobility aircraft plan*. https://wsdot.wa.gov/sites/default/files/2025-06/Advanced-Air-Mobility-Aircraft-Plan-May2025.pdf

World Economic Forum (WEF). (2024). *Advanced Air Mobility: Shaping the future of aviation (AVIATE)*. https://www3.weforum.org/docs/WEF_Advanced_Air_Mobility_2024.pdf

World Economic Forum (WEF). (2025). *Advanced Air Mobility: Paving the way to responsible deployment (AVIATE 2025)*. https://reports.weforum.org/docs/WEF_Advanced_Air_Mobility_2025.pdf

Wing. (2023, August 24). *Wing and Walmart partner: Frisco & Lewisville pilots; service hours*. https://wing.com/news/wing-walmart-partner

Wing. (2025, June 5). *Wing and Walmart announce world's largest drone-delivery expansion*. https://wing.com/news/wing-and-walmart-announce-world-s-largest-drone-delivery-expansion-ever

Yedavalli, P., & Mooberry, J. (2019). *An assessment of public perception of Urban Air Mobility*. Airbus UTM. https://www.airbus.com/sites/g/files/jlcbta136/files/2022-07/Airbus-UTM-public-perception-study-urban-air-mobility.pdf

2 Global Regulatory and Certification Standards for eVTOL Integration

Dimitrios Ziakkas

2.1 INTRODUCTION

Aviation has repeatedly reinvented how societies use the sky—first as exploration, then as global commerce, and now as an everyday mobility layer at urban-regional scales. Advanced Air Mobility (AAM) represents the latest turn in this evolution: a systems-level shift that couples aircraft innovation (electrification, distributed propulsion, and high levels of automation) with new operating concepts, infrastructure, and rulemaking. In contemporary usage, the US Federal Aviation Administration (FAA) frames AAM as the safe, efficient integration of highly automated aircraft into the NAS, emphasising that AAM is not a single technology but a collection of new and emerging technologies applied to aviation transportation (FAA, 2024a, 2024b).

2.2 OVERVIEW OF REGULATORY BODIES

Amin Vafadar, Benjamyn Scott, and Dimitrios Ziakkas

The presented historical context identifies the inflection points that distinguish AAM from prior "flying car" eras and introduces the maturity frameworks and policy milestones that now signal system readiness (Figure 2.1). AAM will only move from demonstration to dependable public service if four layers of governance work in concert: national authorities that create certifiable pathways, regional regulators that align those rules across borders, a global body that anchors common baselines, and an industry organisation that translates policy into day-to-day airline and airport practice. In the present landscape, the FAA in the United States and the European Union Aviation Safety Agency (EASA) in Europe publish the most mature, directly applicable frameworks for eVTOL design and operations. The International Civil Aviation Organization (ICAO) convenes States to harmonise definitions and Standards and Recommended Practices (SARPs), so approvals are portable. The International Air Transport Association (IATA)—not a regulator, but the airline trade body—organises roadmaps and working groups that prepare carriers and airports for integration. Taken together, these four vantage points provide the scaffolding within which AAM can scale (EASA, 2019, 2024a; FAA, 2023a, 2023b; IATA, 2023; ICAO, 2024).

DOI: 10.1201/9781003685555-2

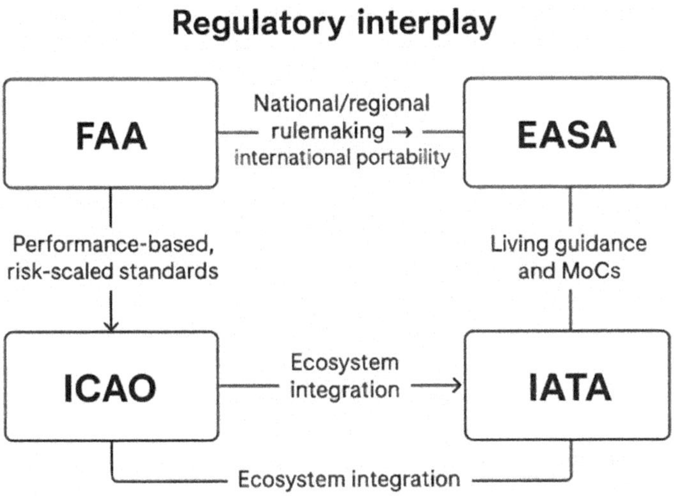

FIGURE 2.1 Historical inflection points from roadable aircraft and urban helicopters to AAM rules and readiness.

- **Federal Aviation Administration (FAA):** The FAA's remit spans airworthiness, operations, personnel licensing, and oversight of the U.S. National Airspace System. Between 2023 and 2025, it set out the pillars that matter most for near-term AAM: the UAM Concept of Operations v2.0, which explains how new entrants integrate with existing air traffic services; the Innovate28 Implementation Plan, which organises concrete actions so that one or more sites can demonstrate at-scale operations by 2028; and the powered-lift final rule—supported by a 10-year Special Federal Aviation Regulation (SFAR)—that formalises pilot certification and operations for eVTOL and wires powered-lift into Parts 61/91/135/136 (FAA, 2023a, 2023b; FAA, 2024a, 2025a). To stabilise vocabulary, the FAA also codified an explicit definition of AAM in the Aeronautical Information Manual (AIM 11-6-1), making clear that AAM is a system for moving people and property using advanced technologies—including electric aircraft—across both controlled and uncontrolled airspace (FAA, 2024b). On certification, eVTOL projects proceed under the powered-lift category via the special-class pathway (14 CFR 21.17(b)), which blends applicable aeroplane and rotorcraft requirements with eVTOL-specific criteria; AC 21.17-4 provides design guidance, while the powered-lift rule/SFAR enables initial Part 135 air-taxi services with performance-based provisions for training, weather minima, altitudes, and noise, aligned to ICAO Annex 1 licensing standards (FAA, 2024c, 2025c). The overarching integration strategy—ConOps v2.0 paired with Innovate28—is deliberately phased: start with corridor-like, piloted operations; add procedures, facilities, and data services; and grow towards denser networks as evidence accrues.

- **European Union Aviation Safety Agency (EASA):** EASA centralises type certification and rulemaking for the EU, while national authorities oversee local approval and enforcement. For eVTOL, EASA pioneered a dedicated airworthiness basis: Special Condition—VTOL (SC-VTOL), first issued in 2019 and consolidated in Issue 2 alongside Easy Access Rules updates in 2024. The framework distinguishes Category Basic (where a controlled emergency landing is the safety objective) from Category Enhanced (continued safe flight and landing), with development assurance levels and reliability targets scaled to passenger risk and operating environment (EASA, 2019, 2024a). Crucially, EASA maintains a living set of Means of Compliance (MoC)—covering structures, propulsion, and energy (EHPS), bird-strike/containment, landing performance, and failure conditions—that have been refined through 2023–2025 consultations. These MoCs make explicit, for example, what constitutes an acceptable demonstration of continued safe flight and landing for Enhanced operations versus controlled emergency landing for Basic (EASA, 2023, 2024b). With Issue 2, EASA also raised the small-category mass ceiling and reiterated the "no single failure catastrophic" design intent—steps that ease convergence with other authorities and accommodate larger multi-motor eVTOL (EASA, 2024c).
- **International Civil Aviation Organization (ICAO):** ICAO does not certify aircraft; its role is to set global SARPs and procedures that States transpose into national rules. For AAM, the task is to harmonise terminology, roles, and performance expectations so that approvals become mutually recognisable and cross-border operations are feasible. In 2024, ICAO convened its first AAM symposium, signalling a structured work programme to align airworthiness, operations, air traffic integration, and facilitation for AAM under a common umbrella (ICAO, 2024). This matters because FAA and EASA can only carry their domestic approvals so far: international validation typically depends on Annex 1 (licensing), Annex 6 (operations), Annex 8 (airworthiness), and related provisions. ICAO's AAM agenda is therefore the venue where early regional practice is translated into globally portable baselines.
- **International Air Transport Association (IATA):** IATA represents airlines and coordinates standards and practices across the commercial aviation value chain. In the AAM context, its New Aviation Ecosystem Roadmap explains how new aircraft types—eVTOL/eSTOL and emerging hydrogen or battery-electric platforms—will intersect with legacy infrastructure, airspace management, and airline operations, emphasising system-centric, digital, and performance-based integration (IATA, 2023). Although many IATA documents are naturally fixed-wing first, the association's workstreams on ATM, safety, and sustainability increasingly address UAS/UTM and other new entrants near airports, shaping airport and airline readiness for energy supply, ground handling, and schedule integration (IATA, 2023, 2025).
- **How these roles interlock:** The four functions are complementary rather than redundant. National and regional rulemaking (FAA powered-lift/SFAR; EASA SC-VTOL; and associated Easy Access Rules) creates binding pathways to certify aircraft and conduct operations; ICAO then crystallises converging practice into SARPs that make bilateral validations and cross-border flights routine rather than exceptional (FAA, 2024e, 2025b;

ICAO, 2024). Both FAA and EASA use performance-based, risk-scaled standards—continued safe flight and landing for higher-exposure operations; training and operational requirements tailored to risk—so innovation is enabled without prescribing a single architecture (EASA, 2019; FAA, 2024). Guidance is intentionally living: EASA's MoCs and the FAA's ACs and notices will keep evolving on propulsion integrity, human-autonomy teaming, vertiport/heliport specifics, and data/cyber obligations as operations mature (EASA, 2023, 2025; FAA, 2023b, 2025c). Finally, ecosystem integration depends on IATA's roadmaps and airline/airport working groups to translate regulation into workable schedules, ground processes, and passenger experience—bridging compliance and commercial viability (IATA, 2023). In short, AAM scales where these layers cohere: rules that can be certified against, international baselines that make approvals travel, and operational playbooks that turn policy into service.

2.3 FAA CERTIFICATION AND OPERATIONAL STANDARDS FOR eVTOLS

Amin Vafadar, Benjamyn Scott, and Dimitrios Ziakkas

In the United States, the FAA has assembled a coherent regulatory architecture for the first decade of eVTOL integration. Four elements are central. First, airworthiness/type certification proceeds in the powered-lift category using the special-class pathway so that novel configurations can be assessed against a composed set of requirements. Second, pilot certification and training are established through permanent amendments to Parts 61/135/142 and a time-bounded SFAR codified as Part 194, intended to stand up an initial cadre while longer-term rules mature. Third, operational provisions tailor Parts 91 and 135 to powered-lift performance and failure characteristics, preserving safety outcomes while enabling early service entry. Fourth, integration guidance—AAM implementation planning and the UAM Concept of Operations—coordinates procedures, facilities (heliports/vertiports), and data services so operations can grow from corridor-like beginnings to denser networks (FAA, 2023a, 2023b, 2024a, 2025a).

2.3.1 Type Certification: Special-Class Powered-Lift

As far as contemporary eVTOL designs do not map neatly onto any single legacy code, the FAA uses 14 CFR § 21.17(b) to compose an appropriate certification basis from Parts 23/25/27/29/33/35 and to add eVTOL-specific criteria or special conditions where needed (FAA, 2025a). The publication of AC 21.17-4, Type Certification—Powered-Lift (July 2025), gives applicants an organised pathway from type design through production and airworthiness certification. The circular situates projects on a safety continuum, pointing to policy statements and acceptable MoC matched to aircraft size, passenger exposure, and intended use (FAA, 2025b). To underpin that continuum, the FAA issued draft policy statements in 2024—PS-AIR-21.17-01 (Safety Continuum for Powered-Lift) and PS-AIR-21.17-02 (Special Class Rotorcraft)—that set risk-scaled objectives (design assurance and reliability targets) without prescribing a single architecture (FAA, 2024b, 2024c). This posture facilitates convergence with international frameworks such as EASA's SC-VTOL while retaining U.S. flexibility.

In practice, an eVTOL programme assembles its § 21.17(b) basis, proposes MoC consistent with AC 21.17-4, and demonstrates that critical functions satisfy the applicable safety objectives along the continuum. Novel features—distributed electric propulsion, transition-mode control laws, and high-voltage energy systems—are addressed through special conditions, targeted analyses, and test evidence acceptable to the FAA (FAA, 2025a, 2025b).

2.3.2 Pilot Certification and Training: Permanent Rules Plus SFAR (Part 194)

The Integration of Powered-Lift: Pilot Certification and Operations final rule (November 2024) adopts permanent amendments and establishes a 10-year SFAR to bootstrap the initial pilot and instructor pipelines while the agency accumulates operational data (FAA, 2024a). Part 194—"SFAR No. 120—Powered-Lift: Pilot Certification and Training; Operations Requirements"—creates temporary alternatives to specified Part 61 provisions, removes early operational barriers, and embeds targeted mitigations for first-of-type service entry; a January 2025 Federal Register correction resolved duplicative SFAR tables (FAA, 2025c). Pilots obtain a powered-lift category rating and, where required, a type rating. Notably, the rule permits approved Part 135 curricula to satisfy defined Part 61 aeronautical knowledge/experience when adding powered-lift and instrument authorisations, thereby accelerating qualification under controlled conditions (FAA, 2024a, pp. 299–304). Even when operators intend to fly VFR revenue services under Part 135, flight crew must hold an instrument powered-lift rating; Part 194 allows integrated training pathways to meet this requirement efficiently (pp. 299–301). Instructor and initial-cadre provisions recognise manufacturer instructor pilots, FAA test pilots, and aviation safety inspectors—allowing, for example, reduced PIC thresholds in supervised experimental training so qualified instructors and check airmen can be fielded quickly (pp. 687–690). Finally, conforming updates to Part 60 acknowledge powered-lift FSTD qualification and cross-reference Part 194, establishing the basis for device-based credit in training and currency (pp. 723–724).

2.3.3 Operational Rules Tailored to Powered-Lift (Parts 91/135 via SFAR)

Early operations are enabled through a performance-based mapping of selected Part 91 and Part 135 provisions to powered-lift characteristics. Where vertical-lift capability and confined-area performance are demonstrated, § 91.119 minimum-safe-altitude relief may mirror helicopter practice, allowing safe operations at lower heights in specified contexts (FAA, 2024a, pp. 417–418, 690). § 91.151(b) fuel-reserve requirements are tailored for vertical-lift flight modes with helicopter-analogous planning where appropriate (p. 690). Applicability tables in § 194.302/§ 194.306 identify which Part 91/135 rules govern powered-lift operations and where SFAR alternatives apply to deliver equivalent safety during the interim period (pp. 687–692). The intent is clear: preserve margins while collecting data that will inform subsequent permanent rulemaking.

2.3.4 Infrastructure and NAS-Integration Touchpoints

On the ground, the FAA's Engineering Brief 105A (27 December 2024) explicitly treats vertiports as a subclass of heliports and supplements AC 150/5390-2D on geometry, markings, parking, and downwash/outwash protections for multi-propulsor aircraft (FAA, 2024d). At the system level, the AAM Innovate28 Implementation Plan and UAM ConOps v2.0 set the phased integration logic—from corridor-like early operations to scaled networks—aligning procedures, facilities, and digital services across priority sites (FAA, 2023a, 2023b).

2.3.5 What This Means for Manufacturers, Operators, and Infrastructure Managers (Near Term)

For manufacturers, the practical anchor is AC 21.17-4: build the § 21.17(b) special-class basis from first principles, place the design on the powered-lift safety continuum, and map every claim to an explicit MoC while tracking the FAA's policy finalisations (FAA, 2025a, 2025b, 2024b). Operators and training organisations should use Part 194 to establish the initial pilot pipeline, integrate FSTD credit where eligible, and deliver the required instrument qualification even for VFR Part 135 service (FAA, 2024a). Airports and ANSPs should align vertiport projects with EB 105A and develop operating concepts against Innovate28/ConOps v2.0, ensuring that facility geometry, procedures, and information exchanges mature together (FAA, 2024d, 2023a, 2023b).

2.4 EASA SC-VTOL CERTIFICATION GUIDELINES

Amin Vafadar, Benjamyn Scott, and Dimitrios Ziakkas

The EASA has created a dedicated, performance-based airworthiness code for small VTOL-capable aircraft—SC-VTOL—to bridge the gap until a permanent certification specification is adopted. The current baseline, Issue 2 (10 June 2024), establishes where the code applies, raises the maximum certified take-off mass to 5,700 kg while retaining a nine-seat limit, clarifies how Basic and Enhanced safety categories differ, and tightens consistency across subparts so that safety objectives are interpreted uniformly (EASA, 2024a). In practical terms, SC-VTOL gives applicants a common, regulator-maintained vocabulary for demonstrating that new architectures can be made as safe as their intended uses require.

2.4.1 Applicability and Scope

SC-VTOL covers person-carrying, unpressurised, small-category VTOL aircraft with a maximum operating speed of VNO ≤250 KCAS (EASA, 2024a, VTOL.2000).

Issue 2 fixes two bounding dimensions: no more than nine passengers and MCTOM ≤5,700 kg (EASA, 2024a, VTOL.2005), a change explicitly flagged in the "Changes introduced" section. Within that envelope, applicants must choose the certification category that matches the intended operational exposure. Enhanced is required for commercial air transport and operations over congested areas and demands continued safe flight and landing; Basic is intended for lower-risk contexts and requires a controlled emergency landing (EASA, 2024a, VTOL.2005). Definitions central to these outcomes—what counts as continued safe flight and landing, and what qualifies as a controlled emergency landing—were refined in Issue 2, and the definition of a vertiport was aligned with the EU's U-space/Innovative Air Mobility package, including references to Implementing Regulation (EU) 2024/1111 (EASA, 2024a, VTOL.2000).

2.4.2 Safety Philosophy and Objectives

The safety philosophy mirrors that of transport-category aircraft, where exposure warrants it. Catastrophic failure conditions must be extremely improbable and must not result from any single failure (EASA, 2024a, VTOL.2510(a)(1)), a principle most consequential for Enhanced operations (EASA, 2019, 2024a). SC-VTOL scales expectations by category: Enhanced demands design assurance, reliability, and system architectures that credibly support continued flight and landing after credible failures; Basic scales those expectations to ensure a controlled emergency landing. EASA requires applicants to use accepted MoC and maintains an explicit process for MoC acceptance, allowing consensus standards to be adopted where they deliver equivalent safety (EASA, 2024a, VTOL.2010). The result is a performance-based code: it specifies what must be achieved, not how to build a particular machine.

2.4.3 Selected Technical Requirements

Across subparts, Issue 2 emphasises operational realism and occupant protection. Subpart B (Flight and performance) requires applicants to establish normal, operational, and limit flight envelopes and to demonstrate take-off and landing performance—including obstacle margins—together with out-of-ground-effect climb capability and credible strategies for icing (certification, detection/avoidance if not certified) (EASA, 2024a, VTOL.2110–2135, 2165–2170). Subpart C (Structures and occupant protection) focuses on durability, aeroelasticity, and emergency conditions. It tightens bird-impact robustness in proportion to passenger capacity; for example, an Enhanced aircraft must sustain a likely bird impact and continue safe flight and landing, while a Basic aircraft with seven or more seats must achieve a controlled emergency landing after such an event (EASA, 2024a, VTOL.2255–2270). Subpart D (Design and construction) addresses flight-control integrity, landing-gear performance, and egress. Issue 2 expands flotation and limited overwater provisions to reflect EU categories for operations on or near water, including floating surfaces (EASA, 2024a, VTOL.2300–2335). Subpart F (Systems and equipment) codifies protections against lightning and high-intensity radiated fields and makes Electrical Wiring Interconnection Systems (EWIS) explicit—an expected evolution for high-power electric architectures. It also clarifies expectations for power generation, energy storage and distribution separation, and aligns flight recorder requirements with EU rules for small VCA (EASA, 2024a, VTOL.2515–2525, 2555).

Subpart G (Flight-crew interface) brings human factors into the certification basis: displays, markings, and operating information must support safe operation, and for Enhanced aircraft, the interface must allow continued safe flight and landing following loss of vision through any one windshield panel (EASA, 2024a, VTOL.2600–2625).

2.4.4 Means of Compliance and Complementary Special Conditions

To improve accessibility and traceability, EASA consolidated SC-VTOL Issue 2 and related MoCs into the first Easy Access Rules for small-category VCA (October 2024), a maintained, machine-readable compendium (EASA, 2024b, 2024c). MoCs continue to evolve: 2024–2025 publications refine propulsion battery thermal-runaway mitigations, recorder specifications, flotation tests, and systems safety topics (e.g., VTOL.2310/2555/2440), with the fifth publication (July 2025) adding and adjusting several MoCs (EASA, 2025a, 2024d). Propulsion-specific safety is addressed under Special Condition E-19 (Electric/Hybrid Propulsion Systems), supported by its own MoCs, such as MOC-EHPS.80 Safety Assessment (proposed 12 March 2025), which sets analysis and assurance expectations for electric/hybrid architectures, including calibration and validation requirements (EASA, 2025b).

2.4.5 What Changed in Issue 2—and Why It Matters?

Issue 2's technical edits have strategic consequences. Raising MCTOM to 5,700 kg (while keeping the nine-seat cap) allows more battery mass or payload and brings the small-category threshold into harmony with the 12,500 lb boundary used elsewhere, improving cross-validation prospects (EASA, 2024a). Making EWIS explicit recognises ageing and fault risks in high-power systems and forces early, design-level mitigations. Clarifications for over-water operations—ditching, emergency flotation, limited-overwater certification routes—directly serve coastal and riverine cities (EASA, 2024a, VTOL.2310/2315). Language in VTOL.2240/2440/2510 was tightened so catastrophic conditions are extremely improbable and not from a single failure, reinforcing Enhanced-level integrity across structures and systems. Finally, the cleanup of design speeds (VMO/MMO) and removal of legacy CS-23 analogies via MoC VTOL.2200 improves internal consistency and aligns better with CS-27/29-style structural reasoning (EASA, 2024a).

2.4.6 Practical Implications for Applicants

For Enhanced operations over cities, the code effectively requires no-single-failure tolerance and credible continued-flight capability. That expectation translates into distributed propulsion with independent power paths, DAL-A treatment for critical functions, robust thermal/fire containment, and physical/functional EWIS separation (EASA, 2024a, 2025a, 2025b). Programme plans should be built around SC-VTOL Issue 2 plus the current MoCs and SC E-19; test campaigns—battery fire and thermal runaway, flotation, recorders—should be locked down against the most recent Easy Access Rules and MoC consultation outcomes (EASA, 2024b, 2025a, 2025b). The 5,700 kg ceiling and harmonised failure-assumption language reduce deltas with the FAA's powered-lift guidance, easing the cross-border validations discussed later (EASA, 2024a).

2.5 HARMONISING CROSS-BORDER CERTIFICATION REQUIREMENTS

Amin Vafadar, Benjamyn Scott, and Dimitrios Ziakkas

eVTOL programmes rarely target a single jurisdiction. Without alignment, each validation becomes a bespoke exercise—duplicating tests, prolonging timelines, and increasing the cost of capital. Regulators have therefore moved beyond ad hoc bilateral discussions towards a structured, multi-authority roadmap that defines common safety objectives, shared documentation, and predictable validation workflows (FAA, 2025a; UK CAA, 2025).

A first expression of this shift is the "NAA Network" playbook. In 2025, the FAA and peer authorities (EASA, the UK CAA, and others) published a roadmap for AAM Aircraft Type Certification, codifying principles that include harmonised type-certification expectations, collaborative multi-authority validation, incremental and risk-based approaches, and AAM-inclusive bilateral instruments (FAA, 2025b; Holland & Knight, 2025). The roadmap points explicitly to FAA AC 21.17-4 (Powered-Lift Type Certification) and to SC-VTOL (as adopted by EASA and the UK CAA) as anchor baselines that should be cross-referenced during validation. The practical aim is simple: reduce the deltas up front, so a type certificate earned in one State can be validated elsewhere with minimal rework.

Convergence is visible in the technical baselines. In the United States, AC 21.17-4 (2025) organises the special-class powered-lift basis around a safety-continuum philosophy and complements the 2024–2025 powered-lift pilot/operations rule and SFAR that integrate powered-lift into Parts 61/91/135/142 (see also industry coverage summarising early OEM use, AIN, 2025). In Europe, SC-VTOL Issue 2 (2024) raised the small-category mass limit, sharpened the Basic versus Enhanced safety outcomes, strengthened no-single-failure language, and continues to iterate MoCs on propulsion, recorders, flotation, and EWIS (EASA, 2024c). In the United Kingdom, the CAA has consulted on UK.SC-VTOL Issue 2, signalling a commitment to remain aligned for cross-border validations. Read together, AC 21.17-4 and SC-VTOL-02 now point to substantially similar safety outcomes—risk-scaled objectives and no-single-failure integrity for urban commercial operations—shrinking the design differences that historically complicated validations.

Validation rides on the machinery of Bilateral Aviation Safety Agreements (BASAs) and their implementing procedures. The FAA-EASA TIP specifies how each side accepts the other's design approvals, data, and continued-airworthiness support; both agencies have pursued risk-based improvements since the Validation Improvement Roadmap and are now extending those instruments so powered-lift/eVTOL fits cleanly within standard pathways envisioned by the NAA Network. Policy signals reinforce this trajectory: in June 2024, the FAA and EASA marked a milestone towards aligned eVTOL policy and promised deeper cooperation on emerging technologies; beyond Europe, the FAA's Declarations of Cooperation with Japan (JCAB) and Korea (KOCA) serve as pathfinders for Asia-Pacific harmonisation (public statements and Reuters coverage, 2024).

Where prescriptive alignment is premature, authorities lean on consensus standards—ASTM, RTCA/EUROCAE—as accepted MoCs, especially for UTM/

Remote ID interfaces, systems safety, recorders, and EWIS. EASA's SC-VTOL portal tracks MoC updates; AC 21.17-4 endorses performance-based acceptance of such standards, letting applicants show equivalent safety while maintaining international acceptability.

For applicants, three practices reduce friction. First, pick a lead authority but plan globally: base the certification plan on AC 21.17-4 (U.S.) or SC-VTOL-02 plus MoCs (EU/UK), and map requirements to the NAA Network's validation expectations. Second, lock deltas early with a structured gap analysis—weight and seats, failure assumptions, recorder and flotation provisions, and EWIS architecture—so the type-certification basis is validation-ready in target markets. Third, pre-coordinate special conditions and MoCs via collaborative validation channels to agree on test witnessing, simulation credit, and data formats once, then reuse. Pair the airworthiness pathway with the appropriate operational rules—the U.S. powered-lift pilot/ops rule and SFAR, or Europe's Innovative Air Mobility rules—so type-design evidence and operating specifications mature together.

2.6 AI-DRIVEN CERTIFICATION AND AUTOMATED SAFETY VALIDATION—DAEDALIAN CASE STUDY

Amin Vafadar, Benjamyn Scott, and Dimitrios Ziakkas

As eVTOL programmes incorporate perception, guidance, and decision-support functions that cannot be hand-engineered deterministically—visual traffic detection, runway/vertiport cueing, obstacle, and wire detection—regulators are formalising how machine-learning constituents can be approved. Classical software assurance (e.g., DO-178C/ED-12C) presumes traceable logic and complete determinism; data-trained models do not fit that mould. The emerging question is not whether AI can be used, but how to scope, train, verify, validate, approve, and monitor ML components so they deliver airworthiness-grade performance within a declared operational design domain (ODD) (EASA, 2023; FAA, 2024e).

EASA's AI Roadmap 2.0 (May 2023) defines a human-centric framework with three building blocks—trustworthiness analysis, AI assurance, human factors, and risk mitigation—and proposes a phased path from Level 1 assistance to more autonomous functions (Figure 2.2 EASA, 2023).

Concept papers translate the roadmap into practicable guidance for Level 1 (and later Level 2) ML applications, that is, functions that assist humans and make bounded decisions under human oversight (EASA, 2021, 2024). EASA has also issued SC-AI-01 (Issue 01, April 2022), a special condition that introduces AI-specific objectives—evidence of explainability, disciplined data management, and robustness—where legacy codes are silent (EASA, 2022). The MLEAP programme extends methods for robustness, stability, scenario-based testing, and runtime monitoring (EASA, 2023, 2024d). In the United States, the FAA's Roadmap for AI Safety Assurance (July–September 2024) sets complementary principles: use existing rules where possible, field incremental use cases, distinguish learned from learning AI, and emphasise runtime assurance. It encourages early projects with lower criticality to build shared experience and identifies research and MoC priorities for perception

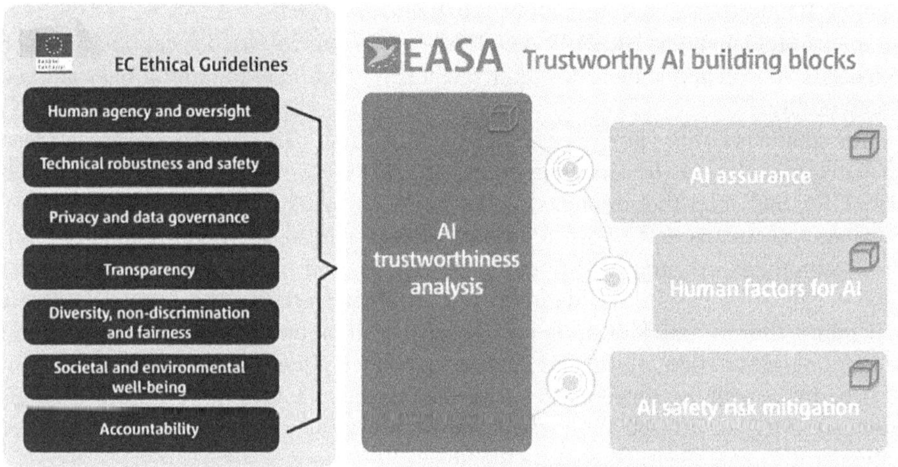

FIGURE 2.2 EASA Ethical Guidelines—Trustworthy AI building blocks (EASA, 2023).

functions (FAA, 2024f). In parallel, EUROCAE WG-114 is developing ED-324 (target 2025), a process standard intended to serve as a harmonised MoC for EASA/FAA policy (EUROCAE, 2025).

Daedalean's collaboration with EASA illustrates how these ideas become certification practice. Beginning in 2019, the partners launched an Innovation Partnership Contract that produced the Concepts of Design Assurance for Neural Networks (CoDANN)—the first aviation-focused blueprint for Learning Assurance (Daedalean & EASA, 2019). CoDANN proposes a "W-shaped" life cycle that augments classical V&V with ML-specific loops: disciplined data specification and curation, training-process assurance, model verification and validation (including generalisation and robustness claims), and post-deployment monitoring with configuration control of the frozen model at entry-into-service (no on-aircraft online learning) (Figure 2.3). Later work (often dubbed CoDANN II) and MLEAP deliverables deepen guidance on distribution shift, ODD definition, scenario coverage, and how error budgets connect to functional hazard assessments (EASA, 2024e).

A concrete use case is Ailumina™ Vista, aimed at visual traffic detection for see-and-avoid support. Public material indicates two active STC programmes—one in CS-23 (with FAA involvement) and one in CS-27 (with EASA)—both initially at DAL-C and grounded in SC-AI-01 for the AI constituent, with concurrent validation anticipated (Daedalean, 2025). The assurance strategy mirrors the regulators' expectations: stringent data design (provenance, labelling quality, and domain coverage), disciplined learning-process control (repeatability, hyperparameter traceability, and training records), and model-level verification (robustness to corruptions and edge cases, traceability to hazards and ODD boundaries). These artefacts are amenable to staged regulator audits and align with the FAA's advice to begin with bounded, lower-criticality applications (Daedalean, 2025; FAA, 2024d).

Automated safety validation rests on scenario-based testing and ODD discipline. Applicants must demonstrate that training/test datasets and simulation campaigns

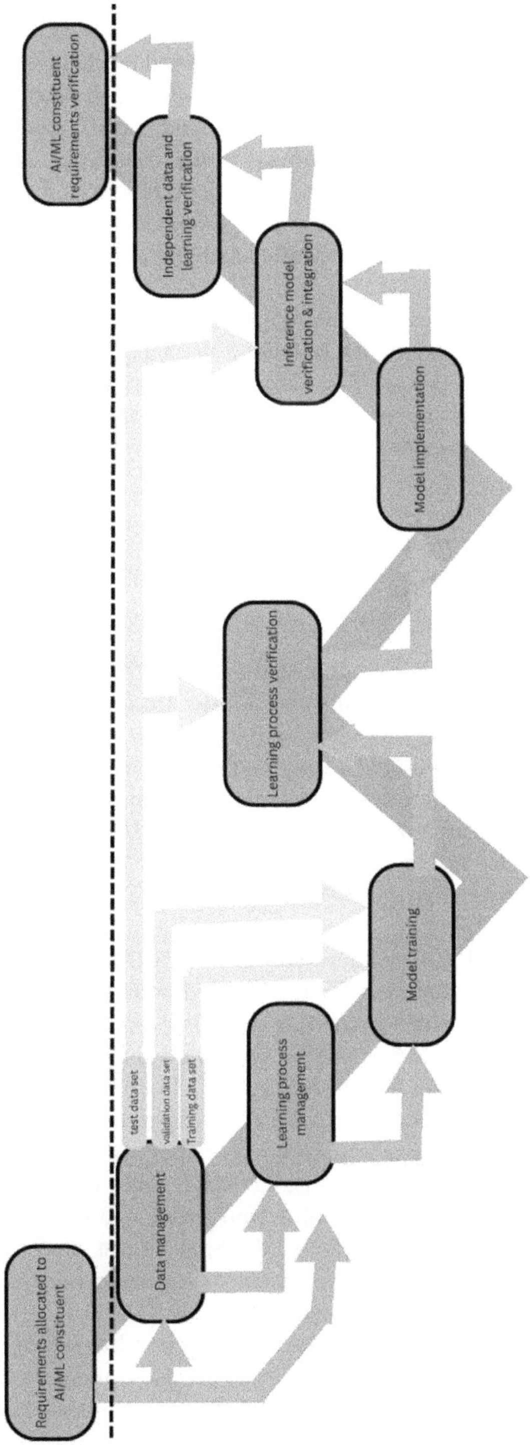

FIGURE 2.3 EASA proposed W model (EASA, 2023).

cover nominal and corner cases, quantify coverage gaps, and bound out-of-distribution behaviour (EASA, 2023, 2024e). CoDANN calls for quantified robustness—to noise, occlusions, compression, and sensor faults—supported by challenge sets and independent hold-outs, with residual error shown to be compatible with system-level safety margins (EASA & Daedalean, 2020). The FAA roadmap stresses runtime assurance: online monitors detect ODD exits, estimate model confidence, and trigger graceful degradation or hand-back to the human or alternate functions. Configuration control of the released model is part of the approval argument. When ED-324 lands, these patterns should crystallise into reusable MoCs, enabling cross-jurisdictional acceptance (EUROCAE, 2025).

For eVTOL teams, the message is pragmatic. Begin with crew-aid functions at DAL-C/B with tightly declared ODDs; build a W-shaped evidence stack where data, learning process, and model are first-class, configuration-controlled artefacts; plan runtime monitors and procedures that support graceful degradation; and track consensus standards so the same evidence base will support cross-border validation (§ 2.5).

2.7 EASA-SOUTHERN EUROPE eVTOL CERTIFICATION PILOT PROJECT

Dimitrios Ziakkas, Benjamyn Scott, and Amin Vafadar

In February 2025, Benidorm (Spain) hosted a U-ELCOME demonstration that exercised Europe's Innovative Air Mobility and U-space framework in a dense coastal city. The event executed the first coordinated urban flight of a pilotless EH216-S eVTOL (no passengers) alongside multiple drones and manned aircraft, while three U-space service providers (USSPs) delivered digital planning, separation, and monitoring. The pilot shows how European law—Regulation (EU) 2021/664—and national oversight (AESA, Spain's CAA) enable staged urban trials that produce evidence for certification and future operational rules (EASA, 2024c; EHang, 2025; ENAIRE, 2025; EUROCONTROL, 2025).

Governance reflected Europe's layered model. EASA set the EU-level airworthiness and operational foundations—SC-VTOL for design and U-space for low-altitude digital services—supplemented by AMC/GM and coordination with SESAR JU and EUROCONTROL (EASA, 2024e). AESA authorised and oversaw the local trial, with ENAIRE integrating activities into the broader ATM environment. EUROCONTROL, as U-ELCOME coordinator, and SESAR 3 JU orchestrated multi-site demonstrations to validate U-space services at scale and their interfaces with ATM (EUROCONTROL, 2025; SESAR JU, 2025). Operationally, ENAIRE and peer USSPs delivered plan management, tracking, strategic/tactical deconfliction, and traffic information, while EHang supplied the aircraft and UPV led the Spanish test cluster. Benidorm's high-rise coastal setting made it a useful stress test for communications, surveillance, urban wind effects, and population-density risk mitigations.

The trial activated the U-space legal scaffolding in a live setting. Regulation (EU) 2021/664 defines U-space airspace, mandatory services, and the roles of USSPs and competent authorities, creating a legal basis for digital, automated separation support at low altitude; EASA's Easy Access Rules (May 2024) consolidate the Regulation with AMC/GM, including cyber requirements that become effective in 2026, and allow Member States to designate U-space and approve USSPs (EASA, 2024e; European Commission, 2021). Although the EH216-S flew without passengers, the exercise sits on the same safety continuum that will bind manned eVTOL operations: SC-VTOL Issue 2 for airworthiness and emerging IAM operational and licensing provisions (Reg. 2024/1111 and 2025 AMC/GM) for pilots and operations. National oversight was visible throughout, with AESA/DGAC approvals and ENAIRE's ATM interface demonstrating that Member States can operationalise EASA's framework in real cities.

Technically, the EH216-S traced an urban profile along the Poniente/Levante seafront while 12 drones executed logistics, surveillance, and rescue simulations, and manned aircraft operated in the vicinity. The combined operation validated multi-domain deconfliction and situational awareness across actors. Three USSPs ran parallel services—flight planning, tracking, strategic deconfliction, tactical geo-fencing, traffic information, and emergency coordination—with an ATM interface to ENAIRE, consistent with U1/U2 capabilities that U-ELCOME aims to mature to TRL 8 across 15 sites in Spain, Italy, and France (EUROCONTROL, 2025). Safety was preserved through segregated volumes, procedural mitigations, and operational limits (no passengers, bounded routes/altitudes, benign weather windows), the point being to collect the operational data needed to refine MoCs and AMC/GM.

What did the pilot prove? First, an end-to-end orchestration of a passenger-class eVTOL (uncrewed in this case) with drones and manned aircraft is feasible in a live city when supported by multiple USSPs (ENAIRE, 2025). Second, the U-space construct can manage mixed low-altitude traffic under national oversight consistent with EASA rules (EASA, 2024c; European Commission, 2021). Third, U-space-ATM interoperability—information exchange and emergency coordination—can be made to work in practice (EUROCONTROL, 2025). It did not yet demonstrate passenger-carrying urban eVTOL commerce, Enhanced-category equivalence in a manned CAT operation, or vertiport-to-vertiport service with energy/turnaround logistics and noise corridor design; those elements will be proven at other sites (e.g., Rome/Fiumicino vertiport trials) and later phases (ADR/Volocopter, 2022).

Three lessons stand out. U-space is the operational backbone for mixed traffic; eVTOL programmes should plan U-space integration artefacts—interfaces, conformance, and resilience—into their safety cases from the start (EUROCONTROL, 2025; European Commission, 2021). Member States will tailor EASA frameworks to local risk pictures; early engagement with the competent authority—hazard workshops and trial CONOPS—de-risks timelines (ENAIRE, 2025). Finally, data capture matters as much as flight: tracking quality, latency, alert rates, human-system interactions, and public-acceptance observations must be curated and aligned with eRules/SESAR reporting so they can feed MoCs and AMC/GM updates (EASA, 2024c; EUROCONTROL, 2025).

For Southern Europe, the near-term implications are concrete. With U-ELCOME momentum and ENAIRE at the table, Spain is positioned to designate U-space corridors for early pilots—port/coastal logistics, tourism shuttles—subject to vertiport readiness and noise/visual-impact studies (ENAIRE, 2025; EUROCONTROL, 2025). Italy's Rome/Fiumicino work created vertiport precedents useful for manned VCA operations as EASA's IAM rules mature; cross-pollination between Rome's infrastructure lessons and Spain's U-space procedures will accelerate cross-border harmonisation

(ADR/Volocopter, 2022; EASA, 2024b). The immediate watchlist includes the rollout of IAM AMC/GM and their adoption into national ops-specs, the maturation of USSP approval pipelines and U-space designations in more Southern-European cities (Valencia, Seville), and the transition from demo to revenue—aligning vertiport standards, noise corridors, public acceptance, and insurance under the EASA/Member-State umbrella (EASA, 2024a).

2.8 CONCLUSION

Dimitrios Ziakkas, Amin Vafadar, and Debra Henneberry

Chapter 2 has traced how AAM is being made real not by a single statute or specification, but by a web of complementary instruments that now hang together with surprising coherence. At the national and regional level, the FAA and EASA have each put durable anchors in place: in the United States, the powered-lift final rule and Part 194 SFAR unlock pilot pipelines and tailor Parts 91/135 to the performance and failure characteristics of eVTOL, while AC 21.17-4 organises the special-class, powered-lift certification basis around a safety-continuum philosophy. In Europe, SC-VTOL Issue 2 establishes a performance-based airworthiness code for small VTOL-capable aircraft, raising the mass ceiling to 5,700 kg (with nine seats), sharpening the Basic versus Enhanced safety outcomes, and making EWIS a first-class design concern for high-power electric architectures. These are not paper exercises: they are the rules, circulars, and MoC that applicants can use today to assemble a certification basis, to propose tests and analyses, and to demonstrate that novel architectures meet risk-scaled objectives.

Two system-level insights follow from this record. First, both authorities have converged—substantively if not yet textually—on performance-based, risk-scaled outcomes: "continued safe flight and landing" for higher-exposure operations; "controlled emergency landing" where exposure is lower; no single failure precipitating a catastrophic condition in the urban, commercial context; and design assurance that scales with consequence. Second, the agencies are deliberately keeping guidance "living": EASA's rolling MoC publications (including propulsion battery thermal runaway, flotation, and recorder topics) and the FAA's evolving policy statements and ACs signal an institutional willingness to absorb evidence and adjust. That stance is not permissive; it is a recognition that credible pathways for new technology require feedback loops.

Harmonisation is the third pillar. The 2025 NAA Network roadmap—explicitly referencing AC 21.17-4 on one side and SC-VTOL on the other—moves cross-border acceptance from hopeful bilateralism to a structured, multi-authority playbook. Read together with existing BASAs and the FAA-EASA TIP, it points to a practical future in which early deltas are identified and burned down before test campaigns begin, test witnessing is shared, simulation credit and data formats are agreed upon once, and type certificates are validated with minimal rework. The policy signals are aligned with this ambition: public commitments to eVTOL cooperation, Easy Access Rules that consolidate SC-VTOL and MoCs into a single, machine-readable body of law, and national consultations (e.g., UK.SC-VTOL) that track European baselines rather than fork them. If applicants reciprocate—by running early gap analyses on weight, seats, failure assumptions, recorders, flotation, and EWIS—validation can become an administrative process, not a new design problem.

A fourth strand concerns AI/ML assurance, which Chapter 2 treated as integral rather than an annex. As perception and decision-support functions move from the lab to operational avionics—visual traffic detection, obstacle, and wire recognition—the tension with classical software assurance becomes unavoidable. Here, EASA's SC-AI01, the AI Roadmap 2.0, and the CoDANN/"W-shaped" life cycle provide the first aviation-specific blueprint for "learning assurance": curate and govern data; control the learning process; verify and validate the trained model for robustness and generalisation within a declared ODD; and monitor the released, frozen model at runtime with clear hand-back pathways. The FAA's AI Safety Assurance roadmap complements this with an incremental strategy that begins with bounded, lower-criticality functions and emphasises runtime assurance. The developing EUROCAE ED-324 process standard promises to translate these ideas into reusable MoC, giving regulators a common language and applicants a portable evidence stack. The practical message for eVTOL teams is simple: treat data, learning process, and model as certifiable artefacts under configuration control, and design runtime monitors as deliberately as you design energy isolation or EWIS separation.

Finally, the chapter grounded these legal and methodological abstractions in the Benidorm U-ELCOME pilot. That trial—an uncrewed EH216-S flying coordinated urban profiles alongside drones and manned aircraft, managed by multiple USSPs under U-space—demonstrated the operational backbone that European AAM will lean on at low altitude: digital planning, strategic and tactical deconfliction, traffic information exchange, and ATM interface under national oversight. It did not (and could not) answer every question—there were no passengers; vertiport-to-vertiport energy logistics and noise corridors remain ahead—but it showed how airworthiness baselines, digital services, and national governance can be orchestrated in a real city, generating the operational data that MoCs and AMC/GM now need.

Taken together, these strands suggest a workable doctrine for near-term AAM:

- Anchor to the right baselines. In the United States, we have to follow the § 21.17(b) basis explicitly against AC 21.17-4 and map every claim to the powered-lift safety continuum; in Europe (and the United Kingdom), we have to frame the programme against SC-VTOL Issue 2, current MoCs, and SC E-19 for propulsion. Early implementation of these steps is highly recommended to avoid delays and last-minute surprises.
- Design for portability. Assume validation, not recertification: adopt data schemas, simulation artefacts, and safety cases that can be read across authorities; pre-coordinate special conditions and MoCs through the NAA Network channels; and document test witnessing and runtime monitors in a way that satisfies both sides of the Atlantic.
- Treat digital as infrastructure. UTM/U-space interoperability, Remote ID, DSS discovery, and vertiport energy management are not add-ons; they are the fabric of daily operations. Build to ASTM F3548/F3411 and national implementations, and expect regulators to scrutinise API-level interoperability and cyber posture as part of the safety argument.

- Human-centric approach. The powered-lift rule and Part 194 are not simply paperwork; they are the means by which high-quality instructors, check airmen, and line pilots are produced and kept current (including instrument privileges, even for VFR revenue service). Human-factors-aware interfaces and procedures remain the first line of defence while automation matures.
- Measure what will matter next. If the end-state is permanent rulemaking beyond SFAR windows, operators must generate the operational data that justify it: dispatch reliability and weather minima performance under corridor procedures; battery and EWIS fault rates; noise footprints versus modelled contours; U-space service-level metrics; training outcomes and FSTD credit efficacy; and, for AI functions, ODD conformance and runtime monitor performance.

The risk, if these practices are not adopted, is predictable: fragmented requirements, duplicated tests, stranded infrastructure, and a public left unconvinced that AAM is transport rather than theatre. The opportunity, if they are, is equally clear: a portable type-certification basis, repeatable operational approvals, and a cadence of evidence that allows regulators to replace interim instruments with durable rules.

REFERENCES

AIN Online. (2025, July 15). *Innovative Air Mobility: New AMC/GM covering VCA operations (pre-flight preparation, fuel/energy management, landing site availability).* https://www.ainonline.com/aviation-news/futureflight/2025-07-15/easa-issues-innovative-air-mobility-rules-covering-evtol

Daedalean. (2019, May 24). *Daedalean and EASA launch Innovation Partnership Contract (IPC): "Concepts of design assurance for neural networks".* https://www.daedalean.ai/tpost/tydicc1kja-daedalean-and-easa-launch-innovation-par

Daedalean. (2025, April). *Ailumina™ Vista: STC pathway and certification status (DAL-C; CS-23/CS-27; SC-AI-01).* https://www.daedalean.ai/certification/ailumina-vista

EHang. (2025, February 28/March 3). *EH216-S conducts first urban eVTOL flight under U-ELCOME in Europe (Benidorm).* https://verticalmag.com/press-releases/ehang-eh216-s-conducts-first-urban-evtol-flight-under-u-elcome-project-in-europe/

ENAIRE. (2025, February 27). *Benidorm hosts the first simultaneous operation in Europe of an air taxi with drones in an urban environment and U-space* [Press release]. https://www.enaire.es/ss/Satellite?blobcol=urldata&blobkey=id&blobtable=MungoBlobs&blobwhere=1488454445585&ssbinary=true

EUROCAE (WG-114). (2025, August 1). *ED-324 (Draft): Process standard for development and certification approval of aeronautical products implementing AI (open consultation).* https://www.eurocae.net/open-consultation-for-ed-324/

EUROCONTROL. (2025). *U-ELCOME project overview: U1/U2 services across 15 locations in Spain, Italy, and France; coordination with EASA and SESAR 3 JU.* https://www.eurocontrol.int/project/u-elcome

European Commission. (2021, April 22). *Commission Implementing Regulation (EU) 2021/664 on a regulatory framework for the U-space.* https://eur-lex.europa.eu/legal-content/EN/TXT/PDF/?uri=CELEX%3A32021R0664

European Union Aviation Safety Agency (EASA). (2019, July 2). *Special condition for small-category VTOL aircraft (SC-VTOL-01).* https://www.easa.europa.eu/sites/default/files/dfu/SC-VTOL-01.pdf

European Union Aviation Safety Agency (EASA). (2020, March 31). *Concepts of Design Assurance for Neural Networks (CoDANN): Public report [with Daedalean]*. https://www.easa.europa.eu/en/document-library/general-publications/concepts-design-assurance-neural-networks-codann

European Union Aviation Safety Agency (EASA). (2021, December 20). *Concept paper: First usable guidance for Level 1 ML applications (Issue 01)*. https://www.easa.europa.eu/en/newsroom-and-events/news/easa-releases-its-concept-paper-first-usable-guidance-level-1-machine-0

European Union Aviation Safety Agency (EASA). (2022, April 19). *Special condition SC-AI-01: Trustworthiness of machine learning-based systems (Issue 01)*. https://www.easa.europa.eu/en/document-library/product-certification-consultations/final-special-condition-sc-ai-01

European Union Aviation Safety Agency (EASA). (2023, May 10). *Artificial Intelligence Roadmap 2.0: A human-centric approach to AI in aviation*. https://www.easa.europa.eu/en/document-library/general-publications/easa-artificial-intelligence-roadmap-20

European Union Aviation Safety Agency (EASA). (2024a, April 10). *Commission implementing regulation (EU) 2024/1111—Establishing requirements for the operation of manned VTOL-capable aircraft [News]*. https://www.easa.europa.eu/en/newsroom-and-events/news/easa-publishes-updated-easy-access-rules-aircrew

European Union Aviation Safety Agency (EASA). (2024b, May 29). *Easy access rules for U-space (Regulation (EU) 2021/664) (consolidated; includes AMC/GM and 2023/203 cyber amendment)*. https://www.easa.europa.eu/en/document-library/easy-access-rules/easy-access-rules-u-space-regulation-eu-2021664

European Union Aviation Safety Agency (EASA). (2024c, June 10). *Special condition for small-category VTOL-capable aircraft—Issue 2 (SC-VTOL-02)*. https://www.easa.europa.eu/en/downloads/139946/en

European Union Aviation Safety Agency (EASA). (2024d, September 24). *Fire/explosion problematics and rulemaking activities: SC-VTOL Issue 2 and MoCs [Conference briefing]*. https://www.fire.tc.faa.gov/pdf/joint/Sep24Meeting/Deletain-0924-EASAFireActivities.pdf

European Union Aviation Safety Agency (EASA). (2024e, November 4). *Easy access rules for small-category VCA (consolidated SC-VTOL + MoCs)*. https://www.easa.europa.eu/en/document-library/easy-access-rules/easy-access-rules-small-category-vca

European Union Aviation Safety Agency (EASA). (2025a, March 12). *Means of compliance with SC E-19 electric/Hybrid propulsion systems—MOC-EHPS.80 safety assessment (proposed)*. https://www.easa.europa.eu/en/downloads/141725/en

European Union Aviation Safety Agency (EASA). (2025b, July 18). *Special condition VTOL—Means of compliance portal (Fifth publication of proposed MoCs)*. https://www.easa.europa.eu/en/document-library/product-certification-consultations/special-condition-vtol

Federal Aviation Administration (FAA). (2023a, April 26). *Urban air mobility concept of operations (ConOps) v2.0*. https://www.faa.gov/sites/faa.gov/files/Urban-Air-Mobility-Concept-of-Operations-2.0.pdf

Federal Aviation Administration (FAA). (2023b, July 18). *Advanced Air Mobility (Innovate28) implementation plan*. https://www.faa.gov/sites/faa.gov/files/AAM-I28-Implementation-Plan.pdf

Federal Aviation Administration (FAA). (2024a, July 11). *Draft policy statement PS-AIR-21.17-02: Special class rotorcraft*. https://www.federalregister.gov/documents/2024/07/11/2024-14489/draft-policy-statement-on-special-class-rotorcraft

Federal Aviation Administration (FAA). (2024b, July 23). *Roadmap for artificial intelligence safety assurance (Version 1)*. https://www.faa.gov/media/82891

Federal Aviation Administration (FAA). (2024c, September 4). *FAA roadmap on AI safety [Briefing deck, REDAC]*. https://www.faa.gov/about/office_org/headquarters_offices/ang/redac/REDAC_508.06_Fall_2024_FAA_Roadmap_on_AI_Safety_09042024

Federal Aviation Administration (FAA). (2024d, September 5). *Aeronautical information manual—change 3: Section 11-6-1 Advanced Air Mobility (AAM).* https://www.faa.gov/air_traffic/publications/media/AIM_Chg_3_dtd_9-5-24.pdf

Federal Aviation Administration (FAA). (2024e, November 21). *Integration of powered-lift: Pilot certification and operations; Miscellaneous amendments related to rotorcraft and airplanes (Final rule).* https://www.federalregister.gov/documents/2024/11/21/2024-24886/integration-of-powered-lift-pilot-certification-and-operations-miscellaneous-amendments-related-to

Federal Aviation Administration (FAA). (2024f, December 27). *Engineering brief 105A: Vertiport design (supplement to AC 150/5390-2D).* https://www.faa.gov/airports/engineering/engineering_briefs/eb_105a_vertiports

Federal Aviation Administration (FAA). (2025a, January 3). *Integration of powered-lift... (Final rule)—Correction.* https://www.federalregister.gov/documents/2025/01/03/2024-30331/integration-of-powered-lift-pilot-certification-and-operations-miscellaneous-amendments-related-to

Federal Aviation Administration (FAA). (2025b, April). *Roadmap for Advanced Air Mobility aircraft type certification.* https://www.faa.gov/air-taxis/NAA-Network-Roadmap-Advanced-AirMobility-Aircraft-Type-Certification-Edition-April2025.pdf

Federal Aviation Administration (FAA). (2025c, July 18). *Advisory circular 21.17-4: Type certification—Powered-lift.* https://drs.faa.gov/browse/AC/doctypeDetails

Holland & Knight. (2025, June 20). *NAA Network releases first roadmap for AAM type certification.* https://www.hklaw.com/en/insights/publications/2025/06/naa-network-releases-first-roadmap-for-advanced-air-mobility

International Air Transport Association (IATA). (2023). *New aviation ecosystem roadmap.* https://www.iata.org/contentassets/c3f5994d248e45cb877774b41e7b3103/new_aviation_ecosystem_roadmap.pdf

International Air Transport Association (IATA). (2025, February 11). *Concept of operations of battery- and hydrogen-powered aircraft at aerodromes (fixed-wing focused; cross-applicability to AAM elements).* https://www.iata.org/globalassets/iata/publications/sustainability/concept-of-operations-of-battery-and-hydrogen-powered-aircraft-at-aerodromes.pdf

International Civil Aviation Organization (ICAO). (2024, September 9–12). *Advanced Air Mobility Symposium (AAM 2024): Global harmonization and interoperability.* https://unitingaviation.com/wp-content/uploads/2024/02/AAM-2024-Programme.pdf

Reuters. (2024, June 13). *FAA, EASA vow to boost collaboration on certification.* https://www.reuters.com/business/aerospace-defense/faa-easa-vow-boost-collaboration-new-airplane-certification-2024-06-13/

SESAR 3 Joint Undertaking. (2025, March). *Benidorm hosts demonstration of urban air taxi and drone operations under U-ELCOME [News].* https://www.sesarju.eu/news/benidorm-hosts-demonstration-urban-air-taxi-and-drone-operations

UK Civil Aviation Authority (CAA). (2025, June 18). *Consultation for special condition UK.SC-VTOL Issue 2.* https://consultations.caa.co.uk/airworthiness-policy-team/consultation-for-special-condition-uk-sc-vtol-issu/

3 Airspace-Integration Framework

From Today's ATM to High-Density AAM

Dimitrios Ziakkas

3.1 INTRODUCTION

Chapter 3 examines how crewed Advanced Air Mobility (AAM) operations can be integrated into contemporary air traffic management and urban infrastructure without destabilizing existing safety, capacity, and community-tolerance equilibria. The argument is deliberately evolutionary: near-term, piloted eVTOL services will fly under today's VFR/IFR and SERA/Part 91/135 paradigms with helicopter-analogous, performance-based provisions; as densities grow, a disciplined layering of digital services—persistent identity, strategic deconfliction, conformance monitoring, and authenticated data exchange—enables predictable networks at very-low-level altitudes. The chapter reconciles U.S. and European pathways by mapping FAA's UAM ConOps v2.0 and Innovate28 work program to the EU's legally binding U-space framework and emerging Innovative Air Mobility (IAM) materials, emphasizing where institutional roles (ANSP/ATS units, U-space service providers, and Common Information Services), procedures (corridors, performance-based navigation, and contingency playbooks), and assurance methods (E-SRM and AMC/GM) converge.

The discussion proceeds in three steps. First, it defines a near-term operating picture for piloted services, including route design, minima, workload management, and the human-automation boundary in terminal/VLL environments. Second, it develops the ground-infrastructure counterpart—vertiports, vertistops, and vertihalls—showing how geometry (e.g., OFV and RD-based pad sizing), energy systems (ED-308-aligned charging), fire/life safety (NFPA 418/855), and digital interfaces must be co-designed to support reliable turnarounds and transparent community footprints. Third, it formalizes the data and governance fabric required to scale: open protocols for identity and intent exchange, service-level expectations for availability and latency, records and audits for oversight, and safety cases that treat predictability and metering not as conveniences but as primary risk controls. Comparative case evidence from FAA trials and early European demonstrations is used to translate concepts into implementable

DOI: 10.1201/9781003685555-3

"city-packages" with measurable KPIs so that regulators, operators, and municipalities can move from sandbox to service with traceable accountability.

3.2 BUILDING THE FUTURE AIRSPACE FRAMEWORK

Amin Vafadar, Fahad ibne Masood, and Dimitrios Ziakkas

Early advanced air-mobility operations will not replace contemporary air traffic management; they will accrete to it. In practice, first-wave, piloted eVTOL services will fly inside familiar VFR/IFR regimes and under existing ATC paradigms, drawing on helicopter-like performance-based provisions and, where helpful, using published routes or corridors to make flows predictable and to support community dialogue. As volumes rise, digitally mediated services—continuous conformance monitoring, strategic deconfliction, authenticated identity, and standardized data exchange—become the binding tissue that allows mixed users to scale. This evolutionary posture mirrors the FAA's UAM ConOps v2.0 and Innovate28 work program, the EU's U-space legal framework and AMC/GM, and the ASTM layer that supplies interoperability and remote-identification performance standards (ASTM, 2021, 2022a; EASA, 2024a; European Commission, 2021; FAA, 2023a, 2023b).

- **The near-term operating picture (piloted eVTOL under current rules):** In the United States, UAM ConOps v2.0 describes a sequencing that privileges continuity over disruption: early, piloted services operate as any other IFR/VFR participant would—inside Class B/C/D/E/G airspace as applicable—under letters of agreement where needed, with procedures and charting that signal expectations to pilots and controllers alike (FAA, 2023a). ConOps also anticipates corridor-like constructs near complex TMAs when they yield safety, predictability, or noise-planning benefits, without implying permanent exclusivity. Innovate28 then ties this concept to place: it concentrates capability at "key sites," aligning certification, operations, infrastructure, and community engagement so that "at-scale by 2028" is a managed outcome rather than an aspiration (FAA, 2023b). The operator's lived reality in this phase is straightforward: air-taxi missions look like helicopter operations with tailored minima and site-specific procedures, while dispatch systems and vertiports furnish reliable conformance data to ANSPs.
- **Digital services and identity: building blocks for scale:** As sorties multiply, order at low altitude depends less on ad-hoc voice and more on shared digital context. In the UAS domain, ASTM F3548-21 defines UAS Service Supplier (USS) interoperability—discovery, intent sharing, and conflict management—across multiple providers. These are the same functions high-density AAM will require, even for crewed fleets (ASTM, 2021). Identity is equally foundational. The FAA's Part 89 Remote ID rule is in force; ASTM F3411-22a specifies broadcast and network Remote ID performance now referenced globally, including by ICAO, for number-space and registration integration (ASTM, 2022b; FAA, 2025). In short, piloted does not mean offline: authenticated identity, standardized intent/conformance exchange, and API-level interoperability are prerequisites for scale (ASTM, 2021, 2022a; FAA, 2023a).

- **The EU pathway: U-space services and the ATM interface:** NASA's integration research anchors these choices: strategic deconfliction, demand-capacity balancing, contingency management, and human-automation teaming have been analyzed in simulation and flight-like trials, lending empirical weight to ConOps assumptions (NASA, 2018). In the early phases, humans still remain the gatekeepers. Pilots, dispatchers, and controllers carry the cognitive load; automation must be transparent and monitorable, with tooling introduced progressively to avoid mode confusion or workload spikes as frequency rises (FAA, 2023a; NASA, 2018).
- **Future steps:** Near-term project managers should advance four workstreams in parallel.

First, publish helicopter-analogous VFR/IFR procedures and LOAs, adding corridors only where they demonstrably improve safety, predictability, or noise performance (FAA, 2023a).

Second, operationalize digital identity and data exchange—Remote ID plus intent/conformance interfaces aligned to ASTM F3411/F3548—so fleets, vertiports, and ANSP/USSP systems interoperate (ASTM, 2021, 2022b; FAA, 2025). Third, stand up a shared operational picture between operators and ANSP/USSP—live trajectories, alerts, and operational intents—to support scalable, mixed operations (EASA, 2024b; FAA, 2023a). Fourth, agree on contingency playbooks and diversion sites in advance and embed triggers in dispatch and ATC workflows so off-nominal events unfold predictably (FAA, 2023a).

To summarize, scaling AAM is not a wholesale reinvention of ATM. It is a disciplined layering of digital services and standardized interfaces atop familiar procedures—with clear governance and human oversight—so that safe, first operations can graduate to dense, predictable networks (Figure 3.1; ASTM, 2021, 2022b; EASA, 2024b; European Commission, 2021; FAA, 2023a).

3.3 INFRASTRUCTURE PLANNING: VERTIPORTS, VERTISTOPS, AND VERTIHALLS

Amin Vafadar, Fahad ibne Masood, and Dimitrios Ziakkas

Ground infrastructure transposes aircraft promise into public service. This section frames typologies (vertiports, vertistops, and vertihalls), siting and urban-fit criteria, safety- and capacity-driven geometry, power/charging, fire and life safety, and the digital interfaces that bind facilities to U-space/UTM and ATC (Figure 3.2). The emphasis is given to European—with PTS concepts and forthcoming vertiport rulemaking—but aligned to the FAA's interim EB 105A and to consensus standards that facilitate harmonization.

Typologies and roles in the network:

- Vertiports are full-service nodes: touchdown and lift-off (TLOF) and final approach and take-off (FATO) areas, passenger processing, energy systems, maintenance/stand space, and digital interfaces.

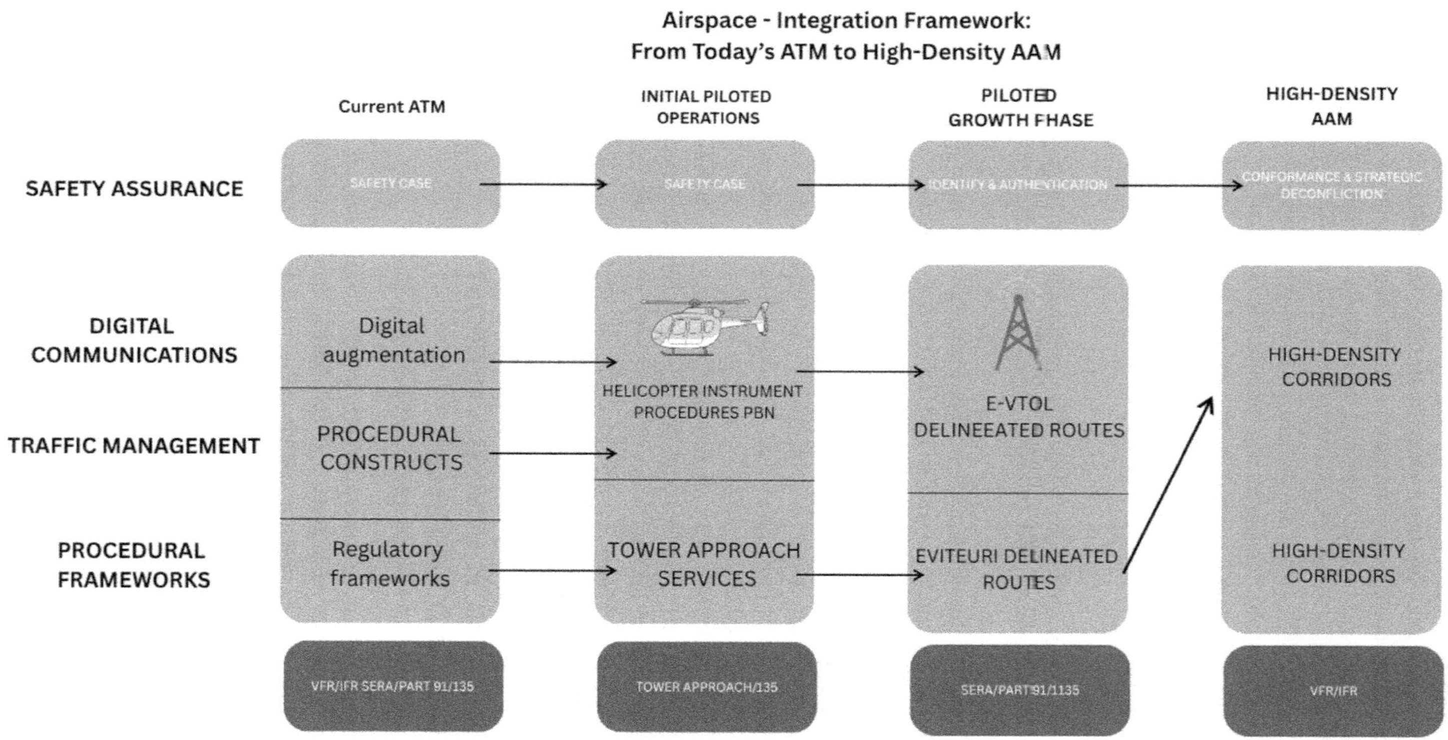

FIGURE 3.1 Airspace integration framework.

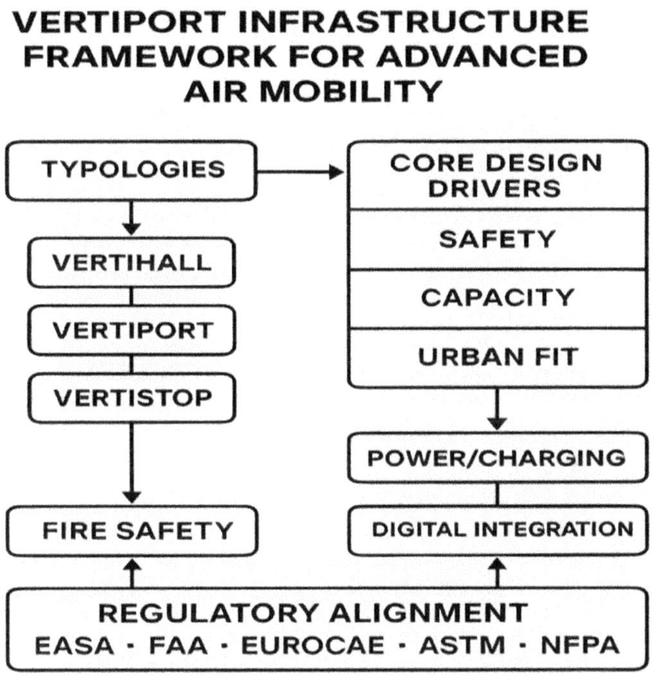

FIGURE 3.2 AAM airspace infrastructure framework presentation.

- Vertistops are minimalist pads for low-throughput or contingency access, with short dwell and no turnbacks.
- Vertihalls denote multi-bay, often enclosed hubs, typically co-located with major intermodal nodes or airports, where weather-protected operations or automated handling can lift throughput.

ASTM F3423/F3423M formalizes vertiport/vertistop distinctions for civil use (ASTM, 2022a). In a tiered network, vertihalls behave as trunk hubs, vertiports as primary urban nodes, and vertistops as end-points and emergency sites. Early deployments gravitate to airport-adjacent and low-controversy locations (industrial/waterfront) to derisk grid interconnection and acceptance.

- **Core design drivers: safety, capacity, urban fit:** EASA's Prototype Technical Design Specifications (PTS) introduced the obstacle-free volume (OFV): a funnel-shaped protected space that supports steep, omnidirectional trajectories suitable for dense fabrics and noise constraints (EASA, 2022). The FAA's EB 105A (Dec 2024) updates U.S. practice by distinguishing vertiports from heliports, adding the "VTL" pad marking, keying controlling dimensions to RD (the smallest circle enclosing propulsors) rather than fuselage span, and formalizing a Downwash/Outwash Caution Area based on empirical surveys (FAA, 2024a). Throughput guidance recognizes that many eVTOL designs hover-taxi rather than ground-taxi; stand separations therefore reflect outwash, wings, and propulsor placement (EASA, 2022; FAA, 2024b).

Additionally, the highest-leverage sites are airport connectors, rail/bus hubs, and park-and-ride nodes that compress door-to-door time. Planning must evidence structural capacity, evacuation routes, heritage/view-corridor preservation, and microclimate effects (urban-canyon winds). EASA's forthcoming Part-IAM AMC/GM stresses landing-site availability, pre-flight preparation, and energy management; those operational expectations should be baked into network design from day one (EASA, 2025d).

- **Regulatory baselines and harmonization:** Under the EASA environment, the PTS is non-binding yet. Moreover, foundational codifying OFV, omnidirectional paths, obstacle assessment, markings/lighting, and feeding RMT.0230 leaning toward a full vertiport rule set (EASA, 2022). U-space EARS (2024a) define the digital layer—authorization, traffic information, and conformance—so USSP-vertiport-ATM interoperate as volumes scale (EASA, 2024b). IAM AMC/GM (2025c) introduces operational materials for manned VTOL-capable aircraft that imply concrete facility procedures (EASA, 2025c).

In the United States (FAA), environment EB 105A is interim, evidence-based guidance pending a performance-based AC; it re-scopes classification, RD-based geometry, VTL markings, parking, and the DCA, and signposts NFPA/IEEE/OSHA references for electrification and safety (FAA, 2024a).

To conclude, ASTM F3423/F3423M (planning/design/establishment), NFPA 418 (2024) for heliports/vertiports, and related NFPA 70/400/855 for electrical and stationary energy storage supply a harmonized baseline.

- **Power, charging, and turnaround:** Turn time and energy throughput govern economics. EUROCAE ED-308 (2023a) sets minimum requirements for VTOL charging infrastructure (EVSE)—installation, commissioning, and O&M—and is compatible with near-term conductive standards now coalescing around CCS hardware (EUROCAE, 2023b). EB 105A directs proponents to NFPA/IEEE/OSHA frameworks and flags grid-quality and cybersecurity considerations (e.g., IEEE-519, NREL guidance) for MW-class charging and on-site storage (FAA, 2024a). Design the energy system as modular and upgradable; isolate ESS/BESS in fire-rated enclosures with detection, suppression, and thermal setbacks; provision load management, DER (PV), and islanding for resilience; and anticipate robotic connectors as standards mature.
- **Fire/life safety and emergency egress:** Rooftop and elevated sites demand redundant egress, smoke/heat management, and joint planning with local responders. NFPA 418 (2024) and NFPA 855 address heliport/vertiport and stationary ESS hazards; EB 105Across-references NFPA 70 (Article 706), OSHA 29 CFR 1910.176, and NREL analyses. Designs should provide fire-rated paths of travel, hydrants/standpipes, foam or water deluge at pads/stands, thermal-runaway containment in battery rooms, clear disconnects/

isolation points, and first-responder access to high-voltage schematics and safe-disable procedures.

- **Environmental and community acceptance:** Acceptance is earned procedurally and acoustically. The OFV supports steep climb/approach segments and omnidirectional routing to reduce exposure footprints and improve obstacle clearance (EASA, 2022). Cities should embed consultation in approvals—transparent noise profiles, hours of operation, and time-bounded trials—site facilities where they displace rather than create surface trips (intermodal nodes), and integrate visually with the skyline (screening and lighting control).
- **Digital integration (U-space/UTM) and ops interfaces:** Facilities must be digitally visible and procedurally integrated. In Europe, U-space mandates network ID, geo-awareness, flight authorization, traffic information, weather (where required), and conformance monitoring through certified USSPs and CIS, enabling pre-tactical slotting, dynamic reroutes, and conflict-free surface/air transitions. Operator manuals should codify paths, missed-approach and diversion logic, and energy-reserve triggers consistent with Part-IAM (EASA, 2024b, 2025d).
- **Practical siting and capacity—what must be true:** Planning approvals (zoning/lease), structural certification, and public consultation come first; safety geometry follows (OFV/FATO/TLOF keyed to RD, setbacks, DCA mapping, obstacle assessment, and markings/lighting). Energy systems must be ED-308-aligned with CCS hardware, and ESS rooms must be built to NFPA standards, backed by grid interconnection studies and DER options. Operations require stands sized for hover-taxi, wet-pad provisions, redundant power/communications, emergency landing/diversion mapping, and interoperability with U-space/ATM. Throughput hinges on a realistic turn-time model and a passenger-flow concept, with risk-based security appropriate to the mission. Finally, training and procedures—first-responder SOPs, ESS incident response, ramp safety, and digital contingency playbooks—close the loop.

3.4 FLIGHT-TESTING PROTOCOLS AND AIRWORTHINESS VALIDATION

Amin Vafadar, Fahad ibne Masood, and Dimitrios Ziakkas

This section lays out a practical, standards-anchored arc from design assurance to flight test in Europe, keyed to SC-VTOL Issue 2 (small-category VCA), the evolving MoCs, and member-State processes for flight conditions and permits to fly. The aim is to align ARP4754B/ARP4761A artefacts with test readiness, envelope expansion, handling-qualities verification, environmental qualification, and evidence generation for certification and early operations (EASA, 2024a).

- **Certification basis and safety objectives:** SC-VTOL Issue 2 defines Enhanced (continued safe flight and landing; needed for CAT and congested areas) and Basic (controlled emergency landing) categories, which drive safety objectives, redundancy, and test points (EASA, 2024b). In 2025,

EASA issued MOC-4 (final, 11 July) and opened MOC-5 (18 July 2025) for consultation, refining handling-qualities methods (including ED-295 references), wind limits, and evidence practices. Programs should treat these as living inputs (EASA, 2025a, 2025b). AMC-20 guidance for lightning and HIRF (AMC 20-136A; AMC 20-158A) and ED-14G/DO-160G environmental qualification complete the baseline for high-voltage architectures (EASA, 2022).

- **Governance: flight conditions, PtF, and FTOM:** Before the first flight, applicants secure flight conditions approval (Part 21 Subpart P; Form 18B annex to Form 37) and a "Permit to Fly" from EASA or the competent authority; the FTOM governs crew competency, test categories, risk assessment, instrumentation, and documentation proportional to program complexity (EASA, 2018). UK policy mirrors EASA to ease cross-border planning.
- **The test and evaluation life cycle:** Under SC-VTOL/MoCs, simulation credit is expected. EUROCAE ED-295 details evidence methods—for example, the Modified Handling Qualities Rating Method (MHQRM)—and defines fidelity/logging expectations; MoCs now cross-reference ED-295 for hover-accuracy and data needs (EASA, 2025a; EUROCAE, 2024a). Conformity of test articles, safety assessments, and Test Readiness Reviews is keyed to approve flight conditions. MOC-3 governs propulsion-battery abuse/containment; EMC/EMI/HIRF/lightning demonstrations follow AMC-20; environmental categories map to ED-14G/DO-160G (EASA, 2023, 2022).

Build-up flights verify hover/low-speed controllability and systems-safety; transition flights characterize control power and failure-management behavior; wind trials meet crosswind/tailwind envelopes; handling qualities follow ED-295 and MOC VTOL.2135 (pilot ratings, tracking tasks, and hover accuracy), including workload and upset recovery (EASA, 2025a; EUROCAE, 2024a). Icing and snow are addressed through ED-314 test or analysis/simulation with validation (EUROCAE, 2024b). Acoustic characterization supports noise statements and local permissions. Data must be configuration-controlled, time-synchronized, and traceable; MOC-4/5 prescribe sampling/accuracy/reporting standards (EASA, 2025a, 2025b).

- **Special topics:**

1. High-voltage propulsion & batteries: MOC-3 emphasizes thermal-runaway initiation, propagation, and containment to protect critical structure, control, and passenger zones.
2. EMC/EMI, HIRF, & lightning: Electric propulsion and distributed inverters increase coupling risks; AMC 20-158A and AMC 20-136A supply acceptable demonstrations, including zoning, shielding, bonding, routing, and aircraft-level test levels consistent with CS-23/25/27/29 references.
3. Software/hardware assurance. DAL-consistent processes (DO-178C/DO-254) within ARP4754B/ARP4761A, with ground/flight verification closing requirements coverage and independence.
4. Member-State example (UK).

For UK-registered aircraft, UK Part 21 Subpart P governs FC and PtF; the CAA or a UK-21J DOA may approve FC. The structure mirrors EASA—application, approval, and PtF issue—with clear validity periods and forms, easing EU-UK demo planning.

3.4.1 Practitioner's Checklist (as Narrative)

Before beginning any test campaign, the applicant should first establish the applicable certification basis under SC-VTOL Issue 2, identifying the correct category of operation and the corresponding Means of Compliance (MoC) to be applied. A comprehensive safety analysis should then be conducted, progressing from the Functional Hazard Assessment (FHA) through Preliminary and System Safety Assessments (PASA and SSA) to define design assurance levels (DAL) and criticality classifications (CCA).

Approval of the Flight Test Operations Manual (FTOM) is an essential milestone, confirming crew qualifications, role assignments, and alignment with the organization's Safety Management System (SMS). Prior to flight, all ground-based evidence for propulsion-battery containment, electromagnetic interference and lightning protection, and environmental qualification must be completed. Flight testing should proceed progressively—beginning with hover and low-speed checks, followed by transition, then cruise phases—validating wind-envelope limits, partial-power or one-engine-inoperative conditions, and overall handling qualities. When relevant, icing or snow-related verifications should also be performed.

Throughout the process, configuration control and instrument calibration must be strictly maintained. Test data should be recorded and reported in accordance with the most recent MoC 4/5 standards, with clearly defined acceptance criteria linked to the underlying SC-VTOL safety objectives (Figure 3.3).

- **Human factors insight:** Integrating representative pilot-handling tasks—such as hover position keeping or precision approaches—early in the simulation-to-flight sequence helps identify workload and mode-awareness issues before high-energy maneuvers. This practice enhances both flight safety and the fidelity of training programs (EASA, 2025a; EUROCAE, 2024a).

3.5 RISK MANAGEMENT AND EMERGENCY PREPAREDNESS

Dimitrios Ziakkas, Fahad ibne Masood, and Amin Vafadar

A credible eVTOL safety posture aligns the operator's SMS (ORO.GEN.200), the aerodrome/vertiport AEP (ADR.OPS.B.005), and RFFS performance (ADR.OPS.B.010), while acknowledging VTOL-specific hazards: high-voltage batteries, distributed propulsion, charging infrastructure, and rooftop/elevated pads. EASA's EARS (Rev. Feb 2025) provide the SMS baseline; ADR Easy Access links the AEP to the aerodrome SMS and external responders.

- **Preparedness and performance:** RFFS must reach any point of each operational surface within three minutes (objective ≤2 minutes under optimum

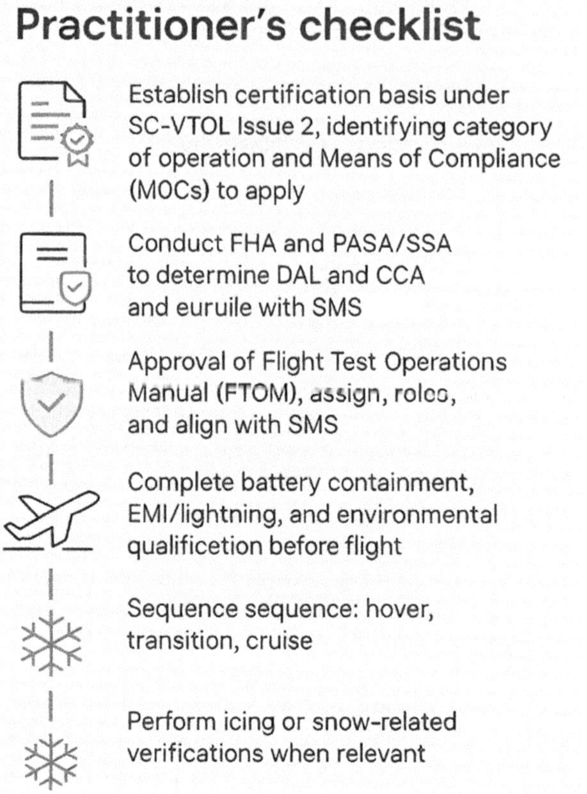

FIGURE 3.3 Practitioner checklist presentation.

conditions); rooftop access constraints and water-adjacent operations modify tactics, staffing, and mutual-aid design. Emergency exercises are cyclical (full-scale ≤24 months; partial exercises in between) with timely remediation of findings. The EU's 2025 ground-handling package clarifies operator responsibilities at the seams with ground-handling providers—turnaround, marshaling, battery handling, and charging—areas central to prevention and response.

- **Hazards and controls:** Thermal runaway requires isolation, rapid detection, smoke management, and reignition monitoring; ED-308 interlocks, signage, and emergency shut-offs must enter the AEP; ED-309 energy-state displays and predictions anchor reserve policy and diversion logic; and ED-307 provides evacuation performance targets that should be validated at representative (including rooftop) sites. Third-party risk controls—pad guarding, downwash assessments, controlled public access—are part of everyday operations.

- **Airspace contingencies:** For Enhanced-category aircraft, operators should publish contingency routes to alternates and emergency sites that honor crowd-density buffers and SC-VTOL definitions of continued safe flight and landing versus controlled emergency landing. Where U-space and crewed ops intersect, operator and vertiport procedures must link to USSPs so emergency descents or reroutes propagate quickly across service layers and ATS.
- **Human-factors note:** Drills that combine energy-state decision-making (ED-309), hover/low-speed handling (ED-295), and rapid egress (ED-307) reveal workload pinch points before real events, particularly on constrained or elevated pads.

3.6 AIR TRAFFIC MANAGEMENT AND U-SPACE INTEGRATION

Amin Vafadar and Dimitrios Ziakkas

Crewed AAM remains inside ATM (SERA/VFR/IFR), unless a Member State designates U-space airspace; in designated volumes, four services are mandatory—network identification, geo-awareness, flight authorization, and traffic information—with optional weather and conformance monitoring where the Airspace Risk Assessment (ARA) so advises (EASA, 2021). Procedures between ATS units and USSPs are required where U-space overlaps controlled airspace. CORUS-XUAM adds a corridor vocabulary—predictable lanes, merges, and metering near TMAs—as a design pattern for busy nodes.

- **Designation and design:** States use the ARA to decide where and how to deploy U-space: entry/exit points, buffers to controlled airspace, vertiport proximity, and cross-border seams (EUROCONTROL, 2023). Corridor architecture then follows: bi-directional lanes with vertical or time-based separation; merge/diverge nodes near vertiports; keep-out volumes; and dynamic narrowing/closure for HEMS/SAR or emergencies via dynamic reconfiguration (EASA, 2021).

CIS provides non-discriminatory static/dynamic data to all USSPs; USSPs deliver the mandated bundle and any State-required additions, with management-system, safety, security, and data-quality obligations that map to ATM/ANS requirements (EASA, 2022). ATS units/ANSPs coordinate procedures with USSPs and exchange data for planning, tactical coordination, and contingencies.

AMC/GM set availability/latency expectations (e.g., ≥99% availability for network identification and traffic information), calibrated to local concepts and ARAs. Open protocols and data-quality monitoring are mandatory; operational records—authorizations, dynamic restrictions, and degradations—are retained (typically ≥5 years) to underpin oversight and re-assessment (EASA, 2022).

- **Planning-to-execution dataflows:** In planning, flight intent flows to USSPs for authorization and strategic deconfliction, with capacity checks against corridor envelopes and State constraints; ATS-USSP coordination resolves

interactions with controlled airspace. In flight, traffic information and, if mandated, conformance monitoring maintain order, while dynamic restrictions (e.g., pop-up HEMS) propagate via CIS/USSP with sufficient anticipation (≥2 minutes) to be effective. Post-flight, audit trails feed safety performance and compliance monitoring.
- **Conflict management and human performance:** Strategic authorization avoids planned intersections; tactical traffic information and ATS-USSP procedures address residual encounters, especially at interfaces. Predictable metering and stable routing are treated as risk controls, not conveniences, because they suppress workload in very-low-level terminal airspace.

Outside EU law, the UK CAA's Airspace Modernization Strategy (CAP 1711) and CAP 722 govern UTM/ATM integration and UAS operations. Conceptual compatibility with EU practice is high; certification/authorization pathways remain distinct (DfT/CAA, 2024; CAP 722 series).

- **Implementation and KPIs:** A two-phase cadence works: Foundational—designate initial U-space volumes in VLL, publish ATS-USSP protocols, and trial a two-vertiport corridor with time-based metering; Advanced—expand to city-pair networks, dynamic reconfiguration, cross-border interoperability, and airport-TMA interfaces. Track authorization success, strategic-deconfliction effectiveness, conformance alerts per 1,000 flights, traffic-info latency, dynamic-restriction reaction time (≥2-minute anticipation met), ATS-USSP availability, and learning-closure time for incidents (EASA, 2022).

3.7 FAA UAM AIRSPACE INTEGRATION CASE STUDY: TRANSFERABLE LESSONS FOR EU/U-SPACE

Amin Vafadar and Dimitrios Ziakkas

The FAA's UAM ConOps v2.0 sketches a phasing from today's ATC to corridor-enabled flows and, ultimately, to an extensible traffic-management (xTM) environment for higher-tempo VLL operations (FAA, 2023a). Innovate28 converts this into a site-based program that coordinates certification, pilot/ops rules, infrastructure, environmental review, and engagement so that one or more "key sites" reach initial operational capability by 2028 (FAA, 2023b). The 2024–2025 powered-lift pilot/ops rule and SFAR supply the human-in-the-loop scaffolding that early services require.

The UAM Airspace Management Demonstration (UAMD) and UTM Pilot Program Phase 2 (UPP2) validated authorization/deconfliction workflows, trajectory exchanges, and corridor-like configurations, with the FAA-industry FIMS interface, Remote ID, and conformance functions proving the value of open, auditable exchanges (FAA, 2021, 2023c). NASA development tests and ATM simulations showed that dozens to >100 air-taxi operations per hour can be integrated near airports using existing ATC tools plus scheduling/metering constructs (NASA, 2023).

- **What the trials revealed:** Corridors are a means, not an end: they stabilize flows and reduce workload near hubs, but access remains ATM-based and dynamic. Distributed services need a common fabric—interfaces, SLAs, and records—to be governable. Human performance is the early bottleneck; schedule discipline and stable routing reduce workload markedly. Furthermore, pilot/ops rules are the practical unlock for moving from demonstration to service. The site-based discipline in Innovate28—concentrating resources where readiness aligns—offers a template for EU Member States designating early U-space corridors and city pilots.

- **Transferable practice:** Implementation should begin with a minimum viable corridor configuration—typically linking two vertiports with defined merge points and time-based metering—operated within existing air traffic control (ATC) structures. In parallel, U-space Service Providers (USSPs) or Common Information Services (CIS) analogues should demonstrate that their systems achieve the expected levels of availability and latency established in previous validation programs such as the UPP2 and Urban Air Mobility Demonstration (UAMD).

In developing the associated safety case, operational predictability must be treated as a primary risk barrier, not simply an efficiency objective. Stable routing, sequencing discipline, and metered entry into constrained airspace directly mitigate pilot and controller workload and therefore form part of the safety architecture itself.

Capacity building should proceed incrementally on a city-by-city basis, with each implementation producing a capability package that integrates airspace design, vertiport interfaces, operator and controller training, community engagement strategies, and environmental management plans. These packages can then be refined and replicated across subsequent metropolitan areas, creating a scalable and standardized deployment framework.

Finally, pilot qualification and operational rules should evolve in parallel with airspace procedures. Early harmonization of training curricula, operating minima, and type-rating requirements ensures that human performance, procedural discipline, and regulatory compliance mature concurrently—reducing integration delays and fostering confidence among authorities, operators, and the public.

3.8 CONCLUSION

Dimitrios Ziakkas and Debra Henneberry

AAM does not require a wholesale reinvention of ATM; it requires rigorously staged augmentation. Piloted eVTOL operations can begin within established rules and procedures if three conditions are met: procedures are codified and charted with helicopter-analogous performance logic; infrastructure is treated as both an airfield and an energy asset with verifiable safety envelopes; and digital services ensure persistent identity, strategic deconfliction, and conformance monitoring across heterogeneous fleets and nodes. The EU's U-space construct and the FAA's ConOps/xTM trajectory represent complementary implementations of the same design thesis: predictable flows, authenticated data, and auditable performance as prerequisites for density. Where they differ institutionally, they converge operationally in the need for corridor discipline near hubs, time-based metering at merge points, and pre-agreed contingency plays that minimize unscripted tactical work.

The implications for practice are concrete. Cities and ANSPs should designate initial corridors and vertiport clusters with service-level agreements for identity, traffic information, and alerting, instrumented from the outset with KPIs—authorization success rates, conformance alerts per thousand flights, information latency, dynamic-restriction reaction time, and emergency-response performance. Operators should align flight-ops platforms, energy management, and training syllabi to these targets, logging evidence into SMS and regulatory records that support continuous improvement and cross-border validation. Standards bodies and authorities should continue to translate trial evidence into accepted MoCs, keeping guidance "living" on issues such as EWIS, battery safety, and digital interoperability. If executed with procedural predictability, transparent data, and community-first siting, the pathway from prototype to public service is not speculative: it is a managed transition that preserves today's safety while making room for tomorrow's networks.

REFERENCES

ASTM International. (2021). *F3548–21: Standard specification for UTM USS interoperability*. West Conshohocken, PA. https://www.astm.org/f3548-21.html

ASTM International. (2022a). *F3411–22a: Standard specification for remote ID and tracking*. West Conshohocken, PA. https://www.astm.org/f3411-22a.html

ASTM International. (2022b). *F3423/F3423M-22: Standard specification for vertiport design*. West Conshohocken, PA. https://www.astm.org/f3423_f3423m-22.html

DfT/CAA. (2024, Oct 22). *Airspace modernisation policy overview*. https://www.gov.uk/government/publications/airspace-modernisation/airspace-modernisation

EUROCAE. (2023a). *ED-308: Guidance on VTOL charging infrastructure*. https://www.eurocae.net/ed-308-guidance-on-vtol-charging-infrastructure/

EUROCAE. (2023b). *ED-309: Guidance on VTOL energy-level information provided to the crew*. https://www.eurocae.net/eurocae-open-consultation-ed-309/

EUROCAE. (2024a). *ED-295: Guidance on VTOL flight-control handling-qualities verification*. https://www.eurocae.net/ed-295-guidance-on-vtol-flight-control-handling-qualities-verification/

EUROCAE. (2024b). *ED-314: Compliance methodologies for VTOL certification in inadvertent icing and snow operation*. https://www.eurocae.net/ed-314-compliance-methodologies-for-vtol-certification-in-inadvertent-icing-and-snow-operation/

EUROCONTROL. (2023). *U-space airspace risk assessment: Method and guidelines (Vol. 1, Ed. 1.0)*. https://www.eurocontrol.int/sites/default/files/2023-04/eurocontrol-u-space-ara-method-guidelines-vol1.pdf

European Commission. (2021, Apr 22). *Commission Implementing Regulation (EU) 2021/664 on a regulatory framework for the U-space*. Official Journal of the European Union. https://eur-lex.europa.eu/legal-content/EN/TXT/PDF/?uri=CELEX%3A32021R0664

European Union Aviation Safety Agency (EASA). (2018). *Flight Test Operations Manual (FTOM) guide*. https://www.easa.europa.eu/sites/default/files/dfu/FTOM%20Guide.pdf

European Union Aviation Safety Agency (EASA). (2021). *Commission Implementing Regulation (EU) 2021/664 on a regulatory framework for the U-space*. https://eur-lex.europa.eu/legal-content/EN/TXT/PDF/?uri=CELEX%3A32021R0664

European Union Aviation Safety Agency (EASA). (2022a). *AMC & GM to Implementing Regulation (EU) 2021/664—Issue 1 (ED Decision 2022/022/R)*. https://www.easa.

europa.eu/sites/default/files/dfu/AMC_and_GM_to_Implementing_Regulation__EU__2021–664_-_Issue_1.pdf

European Union Aviation Safety Agency (EASA). (2022b). *AMC-20 (Amendment 23): General acceptable means of compliance (incl. AMC 20–136A lightning; AMC 20-158A HIRF).* https://www.easa.europa.eu/en/document-library/certification-specifications/amc-20-amendment-23

European Union Aviation Safety Agency (EASA). (2022c, March 24). *Prototype technical design specifications for vertiports.* https://www.easa.europa.eu/en/document-library/general-publications/prototype-technical-design-specifications-vertiports

European Union Aviation Safety Agency (EASA). (2023). *MOC-3 SC-VTOL, Issue 2 (Propulsion batteries—thermal runaway).* https://www.easa.europa.eu/sites/default/files/dfu/MOC-3_SC-VTOL_-_Issue_2_-_21_Jun_2023_-_FINAL.pdf

European Union Aviation Safety Agency (EASA). (2024a). *Special condition for small-category VTOL-capable aircraft (SC-VTOL Issue 2).* https://www.easa.europa.eu/en/downloads/139946/en

European Union Aviation Safety Agency (EASA). (2024b, May 29). *Easy access rules for U-space (consolidated AMC/GM).* https://www.easa.europa.eu/en/document-library/easy-access-rules/easy-access-rules-u-space-regulation-eu-2021664

European Union Aviation Safety Agency (EASA). (2025a). *Safety information bulletin: Managing risks of lithium batteries.* https://www.easa.europa.eu/en/newsroom-and-events/press-releases/easa-publishes-new-recommendations-managing-risks-lithium

European Union Aviation Safety Agency (EASA). (2025b, July 9). *EASA steps up regulatory framework for Innovative Air Mobility (Part-IAM).* https://www.easa.europa.eu/en/newsroom-and-events/news/easa-steps-regulatory-framework-innovative-air-mobility

European Union Aviation Safety Agency (EASA). (2025c, Jul 11). *MOC-4 SC-VTOL Issue 2 (final).* https://www.easa.europa.eu/en/downloads/142242/en

European Union Aviation Safety Agency (EASA). (2025d, July 18). *Special condition VTOL—Fifth publication of proposed MoC (MOC-5 SC-VTOL Issue 1).* https://www.easa.europa.eu/en/document-library/product-certification-consultations/special-condition-vtol

Federal Aviation Administration (FAA). (2021, July 29). *UTM pilot program phase 2: Final report.* https://www.faa.gov/sites/faa.gov/files/uas/research_development/traffic_management/utm_pilot_program/FY20_UPP2_Final_Report.pdf

Federal Aviation Administration (FAA). (2023a, April 26). *Urban Air Mobility (UAM) Concept of operations v2.0.* https://www.faa.gov/sites/faa.gov/files/Urban%20Air%20Mobility%20%28UAM%29%20Concept%20of%20Operations%202.0_0.pdf

Federal Aviation Administration (FAA). (2023b, July 18). *Advanced Air Mobility (AAM) Implementation plan ("Innovate28").* https://www.faa.gov/sites/faa.gov/files/AAM-I28-Implementation-Plan.pdf

Federal Aviation Administration (FAA). (2023c, March 14). *UAM Airspace Management Demonstration (UAMD)—REDAC briefing.* https://www.faa.gov/sites/faa.gov/files/REDAC-NASOps-03142023-508.04-Spring-2023-REDAC-NAS-Ops-UAM-Airspace-Demo.pdf

Federal Aviation Administration (FAA). (2024a, October 22). *Integration of powered-lift: Pilot certification and operations (final rule; SFAR).* https://www.faa.gov/newsroom/integration-powered-lift-pilot-certification-and-operations-miscellaneous-amendments

Federal Aviation Administration (FAA). (2024b, December 27). *Engineering brief 105A: Vertiport design (supplemental to AC 150/5390-2D).* https://www.faa.gov/airports/engineering/engineering_briefs/eb_105a_vertiports

Federal Aviation Administration (FAA). (2025a, January 14). *EB 105A Industry Day (program page).* https://www.faa.gov/airports/engineering/engineering_briefs

Federal Aviation Administration (FAA). (2025b, March 19). *Remote identification of drones: Part 89 resources.* https://www.faa.gov/uas/getting_started/remote_id

Federal Aviation Administration (FAA). (2025c, May 21). *Emerging entrants bulletin 25-02: Testing and demonstrations on airports (AGVS).* https://www.faa.gov/airports/new_entrants/bulletins/25_02

National Aeronautics and Space Administration (NASA). (2018). *Urban air mobility airspace integration concepts and considerations.* https://human-factors.arc.nasa.gov/publications/20180005218.pdf

National Aeronautics and Space Administration (NASA). (2023). *AAM National Campaign—Development test executive summary (acoustics & community).* https://ntrs.nasa.gov/api/citations/20230014300/downloads/AAM-NC-054-001_NC_Exec_Summ_Dev_Testing_V1.pdf

National Fire Protection Association (NFPA). (2024). *NFPA 418: Standard for heliports and vertiports* (2024 ed.). Quincy, MA. https://link.nfpa.org/all-publications/418/2024

4 Human Factors, Training, and Licensing in eVTOL Operations

Dimitrios Ziakkas, Debra Henneberry, and Amin Vafadar

4.1 INTRODUCTION

Advanced Air Mobility (AAM) will scale only as fast as its people, procedures, and interfaces allow. While novel airframes and electric propulsion have attracted most of the public attention, the safety margin in early service will be won or lost in the human domain: how pilots are selected and licensed; how they are trained, checked, and supported in high-tempo urban environments; and how their cockpit tools—automation, displays, and alerts—shape workload, attention, and decision-making. eVTOL operations concentrate task density in short cycles and at low altitudes near obstacles, with frequent transitions between hover, conversion, and wing-borne flight. That combination compresses the time available for diagnosis and response and places a premium on disciplined energy management, mode awareness, and conformance to procedures that themselves are still maturing.

Chapter 4 brings those strands together. It first compares the two leading licensing pathways now available to pilots—the Federal Aviation Administration's powered-lift category and associated Special Federal Aviation Regulation (SFAR) in the United States, and the European Union Aviation Safety Agency's type-rating approach within the Special Condition for VTOL (SC-VTOL) and the Innovative Air Mobility (IAM) aircrew package in Europe (EASA, 2019, 2024a, 2024b; FAA, 2024a). It then argues that competency-based training and assessment (CBTA), as elaborated by ICAO and increasingly reflected in both FAA and EASA guidance, provides the most resilient instructional scaffold for a heterogeneous pilot population and a heterogeneous fleet (EASA, 2023; ICAO, 2023). Because many eVTOL programmes will rely on simulation from the outset—owing to limited dual-control aircraft, the need to rehearse rare but consequential events, and the cost of high-cycle training—we examine the role of high-fidelity devices, AI-assisted debriefing, and immersive technologies in accelerating proficiency while maintaining standardisation (CAE, 2023; FAA, 2024a). The chapter also treats cockpit ergonomics and human-machine interface (HMI) design as a safety system in its own right, connecting SC-VTOL handling-qualities objectives and EUROCAE human-factors guidance with practical patterns—unified controls, envelope protections, energy-state awareness—that

DOI: 10.1201/9781003685555-4

reduce cognitive friction in single-pilot urban operations (EASA, 2024a; EUROCAE, 2021; FAA, 2020). A short case study from Southern Europe illustrates how national systems can adapt licensing, training, and infrastructure to regional conditions, and the concluding section surfaces implications for regulators, operators, and training organisations as early services evolve into routine transport.

4.2 LICENSING PATHWAYS FOR AAM-eVTOL PILOTS: FAA AND EASA APPROACHES

Amin Vafadar, Jim Ives, and Dimitrios Ziakkas

The FAA has elected to recognise eVTOL as a distinct category—powered-lift—thereby providing a complete licensing architecture that runs from ab initio through airline-transport privileges. The November 2024 final rule (SFAR No. 120) integrates powered-lift into Parts 61, 91, 135, and related provisions, creating powered-lift category ratings, type ratings per model, and a powered-lift instrument rating. It authorises commercial on-demand operations under Part 135 and clarifies the training, experience, and checking landscape for the first decade of service via a time-bounded SFAR that complements the permanent amendments (FAA, 2024a, 2024b). In practice, transitioning aeroplane and helicopter pilots can receive targeted credit for prior experience while demonstrating competence in vertical flight, conversion, and energy-limited abnormal procedures. The rule also anticipates practical constraints of early fleets—some single-control prototypes, limited trainer availability—by allowing defined simulator credit and, under safeguards, certain single-control training configurations. The result is a pathway that is both categorical—powered-lift is not a subspecies of aeroplane or rotorcraft—and pragmatic, in that it enables operators to stand up crews under a unified standard while detailed curricula and device qualification mature (FAA, 2024a).

- **EASA's VTOL type-rating framework:** EASA has taken a different route. Rather than create a separate license category at the outset, it leverages existing licences and adds a VTOL type rating. Under SC-VTOL, updated in 2024 to refine safety objectives and raise the small-category mass ceiling, aircraft airworthiness is addressed through a performance-based code, while the IAM package updates the aircrew rules so that commercial aeroplane [CPL (A)] and helicopter [CPL (H)] holders can add a VTOL type rating after an approved course and skill test (EASA, 2019, 2024a, 2024b). Early commercial services are framed conservatively—initially day VFR with progressive expansion to night and IFR as evidence accumulates—yet the training syllabi are designed with the aircraft and its automation level in mind, in close collaboration between manufacturers and Approved Training Organisations (Volocopter GmbH, 2025). This approach trades categorical novelty for speed of implementation: by drawing from the existing professional pilot pool, Member States can field the first crews while EASA continues to evolve licensing constructs for a longer-term VTOL license.
- **Comparative analysis and future pathways:** The two regimes reflect different design philosophies. The FAA emphasises categorical clarity and

TABLE 4.1
Comparative licensing pathways for AAM-eVTOL pilots (FAA vs. EASA)

Key Element	FAA/U.S. Perspective	EASA/EU Perspective	Implications for Industry
Regulatory Basis	SFAR No. 120 establishing powered-lift as new airman category.	SC-VTOL and Aircrew IAM package allow type ratings on existing CPL (A/H).	Divergent philosophies: U.S. categorical vs. EU type-rating expedience.
License Structure	Separate powered-lift category ratings with instrument privileges.	VTOL type rating added to CPL/ATPL; Basic (VFR) and Enhanced (IFR).	FAA allows ab initio entry; EASA leverages current pilot base.
Training Architecture	Simulator credits and limited single-control aircraft accepted.	ATO training developed with OEMs using competency alignment.	Simulator-based curricula ensure early proficiency.
Operational Scope	On-demand/charter (Part 135); scheduled Part 121 excluded.	Day VFR CAT initially; IFR/night later.	Focus on early air-taxi and shuttle markets.
Future Trajectory	Full licensing framework for ab initio pilots.	Prospective standalone VTOL license in IAM evolution.	Convergence towards global pilot certification.

long-run scalability: powered-lift is its own thing, and the rulebook, ratings, and simulator credits say so. EASA emphasises near-term feasibility: repurpose existing licenses through type ratings, then layer in VTOL-specific competencies and operational extensions as the system proves itself (EASA, 2024a, 2024b; FAA, 2024a). Neither approach reduces the retraining burden for transition pilots; it simply organises it differently. Aeroplane pilots must master hover, confined-area cues, and low-energy go-around logic; helicopter pilots must become comfortable with wing-borne aerodynamics, higher-speed IFR flows, and energy-managed descents typical of electric propulsion. Both pathways anticipate CBTA-style syllabi and heavy use of simulation to knit those competency gaps efficiently. Looking forward, the FAA's framework already admits ab initio powered-lift training, though its practicality depends on dual-control training articles and device availability. EASA has signalled that a stand-alone VTOL license could follow once initial operations stabilise and a data-rich basis exists for a complete ab initio pathway (EASA, 2024b). In either case, equivalence of safety with legacy aviation remains the target (Table 4.1). Table 4.1 summarises—compares the existing licensing pathways in the global aviation ecosystem.

4.3 COMPETENCY-BASED TRAINING AND ASSESSMENT (CBTA)

Amin Vafadar, Jim Ives, and Dimitrios Ziakkas

CBTA replaces time-served logic with performance evidence. The framework, consolidated by ICAO and increasingly reflected in EASA and FAA practices, defines target competencies—ranging from flight-path management and application of knowledge to communication, teamwork, decision-making, and workload management—and asks instructors to evaluate observable behaviours across realistic scenarios rather than count hours in manoeuvre silos (EASA, 2023; ICAO, 2023). That stance is well-suited to eVTOL because the pilot cohort will be mixed. An experienced helicopter captain may need only modest practice to attain precise hover-position keeping but more time to internalise wing-borne energy management and instrument procedures; an aeroplane first officer will invert that profile. CBTA enables both to reach a common standard without unnecessary duplication, while also making room for ab initio candidates as dedicated eVTOL training pipelines appear (Table 4.2).

- **Regulatory endorsement and oversight implications:** EASA has long allowed competency-based pathways in instrument training and is now extending the same philosophy to VTOL operations, emphasising that training and checking should be demonstrably tied to the operational profile of the specific type (EASA, 2024b). The FAA's powered-lift SFAR does not rename CBTA explicitly, but it permits alternate pathways, recognises simulator credit, and points operators towards performance-based outcomes that are CBTA-consistent (FAA, 2024a). Both authorities will need oversight models that assure comparability across providers without re-imposing rigid hour minima by another name. That implies calibrated instructor qualification, common scenario libraries, and data-informed standardisation across training centres.
- **Implementation in practice:** Industry is already building the scaffolding. High-fidelity devices fused with analytics platforms make it possible to

TABLE 4.2
Competency-based training and assessment (CBTA) for eVTOL operations

Key Element	FAA/U.S. Perspective	EASA/EU Perspective	Implications for Industry
Training Philosophy	SFAR 120 allows flexible, outcome-based structure.	Performance-based CBTA embedded in Easy Access Rules.	Shift from hours to competencies improves oversight.
Competency Model	ICAO Doc 9995 competencies guide training.	Adopted in EASA Evidence-based Training framework.	Global harmonisation simplifies recognition.
Assessment Mechanism	Instructor discretion plus simulator analytics.	Observable behaviour matrix with adaptive feedback.	Real-time data personalises learning.
Implementation Status	CBTA trials via training centres and OEMs.	Operational guidance for airlines expanding to VTOL.	Shared datasets validate CBTA for AAM.
Future Focus	Validation during the 2024–2034 SFAR period.	Formal CBTA integration in future VTOL licensing.	Institutional embedding across the training ecosystem.

track decision quality, control strategy, and error recovery across repeated urban cycles and to adapt the next session accordingly (CAE, 2023). In early operations, the most discriminating test of competence may be sustained performance over multiple short sectors—climb, convert, cruise a few minutes, descend, de-convert, and land—rather than single-event excellence. CBTA shines when competence must be demonstrated as a rhythm rather than a one-off, because it privileges stable behaviours under realistic tempo.

- **Benefits, constraints, and the path to institutionalisation:** For operators, CBTA promises efficiency, because training time is not wasted on already-mastered skills; scalability, because standardised behaviours can be assessed in distributed networks; and safety, because proficiency is evidenced in realistic sequences. The constraints are familiar: instructors need support to assess consistently; regulators need audit tools that respect variability while guarding against drift; and organisations must invest in data governance so that training analytics remain explainable and auditable. The SFAR's ten-year window in the United States and the IAM roll-out in Europe provide a live-fire period in which CBTA methods can be proven, documented, and converted into enduring policy (EASA, 2024b; FAA, 2024a).

4.4 AI-POWERED SIMULATION AND ADAPTIVE LEARNING

Amin Vafadar, Jim Ives, and Dimitrios Ziakkas

In eVTOL programmes, simulation is not an adjunct but the backbone. Limited trainer fleets, the need to rehearse rare but safety-critical events—battery thermal events, degraded mode transitions, and urban missed-approach contingencies—and the economics of short-sector training make devices indispensable from ab initio through recurrent training. Regulators acknowledge this reality: the powered-lift SFAR explicitly credits simulator time within defined envelopes, and EASA encourages ATOs to validate devices against manufacturer flight-test data so that handling-qualities and automation behaviours are reproduced faithfully (FAA, 2024a; EASA, 2024b).

- **Adaptive learning and instructor augmentation:** The addition of AI analytics changes how devices are used. Performance telemetry captured at millisecond resolution can be turned into session-by-session diagnostics—where workload spiked, how mode transitions were managed, whether scan patterns degraded under traffic or weather loading—and it can inform personalised progression without sacrificing standardisation (CAE, 2023). Properly implemented, such tools make instructors more—not less—central: they free human coaches to focus on judgement and resilience while the system flags patterns worthy of attention. The regulatory question is not whether to allow such tools, but how to validate them: models must be benchmarked against human-rated outcomes; explainability and data lineage must be auditable; and credit policy must remain conservative until validation is mature (Table 4.3).
- **Immersive and mixed-reality modalities:** Virtual, augmented, and mixed-reality systems complement fixed-base simulators by bringing cockpit

TABLE 4.3
AI-powered simulators and adaptive learning in eVTOL training

Key Element	FAA/U.S. Perspective	EASA/EU Perspective	Implications for Industry
Simulation Mandate	Simulator credit authorised for powered-lift training.	ATOs are encouraged to use validated simulators with OEM data.	Simulator is the primary medium for ab initio and conversion.
AI Integration	ML algorithms for performance and scenario adaptation.	Adaptive learning in EASA's innovation programmes (MLEAP).	AI reduces instructor bias and improves standardisation.
Immersive Technologies	AR/VR/MR for portable, high-fidelity training.	Mixed reality integrated into the IAM curriculum.	Distributed training lowers cost and increases access
Oversight and Validation	FAA to define credit equivalence for AI simulators.	EASA is developing MoCs for AI and CBTA evaluation.	Regulatory clarity is crucial for AI-based scaling.
Strategic Outcome	Builds safety analytics dataset for FAA policy.	Improves continuous training standards.	Human-AI synergy enhances resilience.

familiarisation and procedure rehearsal to the point of need—vertiports, operator bases, and university labs. These devices are particularly effective for rehearsing vertiport surface flows, rooftop approaches with complex sight cues, and multi-aircraft gate turns. Several manufacturers and training houses are co-developing such systems so that cockpit geometry, control laws, and energy-state presentations align with certified configurations (Garmin, 2022; Honeywell, 2023; Joby Aviation, 2023; Lilium GmbH, 2023; Volocopter GmbH, 2025).

- **Benefits and challenges in an operational system:** The benefits are direct: throughput scales without a commensurate increase in training aircraft; rare events can be practised safely; and geographically distributed networks can uphold a common standard. The challenges are equally concrete: device qualification must keep pace with rapid software and automation updates; AI-based assessors must be treated as advisory until their equivalence to human judgement is demonstrated; and training data must be protected as safety-critical records, not exhaust from a learning tool. None of these constraints is unique to AAM, but the tempo of early eVTOL operations and the novelty of the interfaces make them urgent (CAE, 2023; NASA, 2023).

4.5 CASE STUDY: SOUTHERN EUROPE—eVTOL TRAINING ADAPTATIONS IN GREECE

Dimitrios Ziakkas, Konstantinos Pechlivanis, and Amin Vafadar

Greece provides a useful laboratory for eVTOL training and human-factors integration because geography and demand align with AAM's strengths. A network of inhabited

TABLE 4.4
Case study: Greece/Regional Training Adaptations for eVTOL operations

Key Element	FAA/U.S. Perspective	EASA/EU Perspective	Implications for Industry
Operational Drivers	FAA pathway is validated for Greek pilots (Beta Alia-250).	HCAA aligns with the EASA IAM rollout.	Model of FAA-EASA cooperation in practice.
Industry Partnerships	FAA-approved dual-seat training.	EASA validation after type certification.	Transatlantic synergy speeds adoption.
Training Context	Simulated high-wind and confined approaches.	CBTA for microclimates and urban vertiports.	Localised training increases safety acceptance.
Regulatory Adaptation	FAA license credits are transferable to EASA.	IAM phased rollout: Day VFR → IFR.	Template for small-state EU implementation.
Strategic Outcome	FAA innovations validated regionally.	EASA integration via U-space corridors.	Greece as Southern Europe's AAM testbed.

islands, strong seasonal traffic, and variable micro-weather create short sectors with frequent take-offs and landings, complex approach cues, and exacting energy-reserve management. Early industry partnerships have been structured to bridge transatlantic regulatory pathways and accelerate crew readiness while European VTOL approvals mature. In this model, domestic operators develop their first cadre through programmes validated under FAA rules—thereby leveraging device credit, dual-seat trainer availability, and a defined powered-lift syllabus—before converting crews to European type ratings once EASA certification is secured for the specific aircraft (Beta Technologies, 2024; EASA, 2024b; FAA, 2024a). The research proposal for the Hellenic Civil Aviation Authority's engagement with EASA on the IAM package will allow initial commercial services to begin under day VFR with a clear path to night and IFR as operational evidence accumulates, while cities and port authorities prepare vertiport procedures that dovetail with emerging U-space constructs for drones. Academic partners contribute by embedding CBTA and AI-assisted simulation modules into university and academy programmes so that a broader workforce—pilots, dispatchers, and vertiport managers—acquires a common vocabulary and toolset (OECD, 2023). The lesson is that small states with demanding operational contexts can become early movers by aligning licensing transitions, training technology, and city-by-city procedures in a coherent package (Table 4.4).

4.6 CONCLUSION

Dimitrios Ziakkas, Debra Henneberry, and Amin Vafadar

Human performance is the through-line that connects certification, training, cockpit design, and operational oversight in eVTOL services. The FAA and EASA have chosen different regulatory entry points—categorical powered-lift versus type-rating within the legacy license families—but both converge on the same safety proposition: that pilots will achieve and maintain competencies demonstrably equivalent to those in conventional commercial aviation, even as their aircraft and missions differ. CBTA provides the pedagogical mechanism to deliver that proposition across diverse pilot backgrounds, and simulation—supported by AI-enabled analytics and immersive modalities—provides the scalable infrastructure to implement it. Cockpit ergonomics and HMI choices then translate policy and pedagogy into practice; when unified control logics, robust energy-state awareness, disciplined alerting, and envelope protections are present, they reduce mode confusion and free cognitive bandwidth for the tasks that cannot be automated: anticipation, communication, and judgement.

For regulators, the near-term task is to continue converting interim frameworks into durable, auditable regimes without strangling innovation—codifying device credit where validated, institutionalising CBTA with calibrated oversight, and aligning human-factors expectations across airworthiness and operations (EASA, 2024a, 2024b; FAA, 2024a). For operators and training organisations, the imperative is to treat data as a safety asset: training analytics, line-operations monitoring, and debrief artefacts should feed Safety Management Systems and inform continuous syllabus improvement. For manufacturers, design is a safety control: the most effective training syllabus is the one made unnecessary by an interface that anticipates human limits. As early city pairs move from demonstration to public service, the systems that keep the pilot at the centre—licensing that recognises competencies, training that measures them, devices that build them, and cockpits that respect them—will determine whether AAM's promise becomes a dependable part of everyday mobility.

REFERENCES

Beta Technologies. (2024). *Alia-250 aircraft overview and training programme*. https://www.beta.team/aircraft/alia-250

CAE. (2023). *Advanced Air Mobility training white paper*. https://www.cae.com/media/documents/Civil_Aviation/whitepapers/cae-aam-training.pdf

EASA. (2019). *Special Condition VTOL—Issue 1: Vertical take-off and landing (VTOL) aircraft*. European Union Aviation Safety Agency. https://www.easa.europa.eu/en/document-library/special-conditions/special-condition-vtol-issue-1

EASA. (2023). *Evidence-based training and CBTA guidance*. European Union Aviation Safety Agency. https://www.easa.europa.eu/en/domains/aircrew/training

EASA. (2024a). *Special condition VTOL—Issue 2*. European Union Aviation Safety Agency. https://www.easa.europa.eu/en/document-library/special-conditions/special-condition-vtol-issue-2

EASA. (2024b, December 11). *Updated Easy Access Rules for Aircrew (Innovative Air Mobility package)*. European Union Aviation Safety Agency—Newsroom. https://www.easa.europa.eu/en/newsroom-and-events/news/easa-publishes-updated-easy-access-rules-aircrew

EUROCAE. (2021). *ED-258: Human factors guidelines for VTOL aircraft; ED-295: HMI assessment for VTOL.EUROCAE*. https://www.eurocae.net/

FAA. (2020). *Human factors design standard*. Federal Aviation Administration. https://www.faa.gov/regulations_policies/handbooks_manuals/aviation/human-factors-design-standard

FAA. (2024a, November 21). *Integration of powered-lift: Pilot certification and operations; Final rule (SFAR No. 120)*. Federal Register, 89(223), 92199–92386. https://www.federalregister.gov/documents/2024/11/21/2024-24886/integration-of-powered-lift-pilot-certification-and-operations-miscellaneous-amendments-related-to

FAA. (2024b). *Powered-Lift SFAR—Frequently asked questions*. https://www.faa.gov/air-taxis/FAQ

Garmin. (2022). *G3000 integrated flight deck overview*. Garmin Aviation. https://www.garmin.com/en-US/aviation/

Honeywell. (2023). *Avionics solutions for Urban Air Mobility*. Honeywell Aerospace. https://aerospace.honeywell.com/en/learn/products/urban-air-mobility

ICAO. (2023). *Manual on competency-based training and assessment* (Doc 9995, 2nd ed.). International Civil Aviation Organization. https://store.icao.int/en/manual-on-competency-based-training-and-assessment-doc-9995

Joby Aviation. (2023). *Joby aircraft overview: Flight deck features*. https://www.jobyaviation.com/

Lilium GmbH. (2023). *Lilium Jet cockpit and avionics development*. https://lilium.com/

NASA. (2023). *Human-in-the-loop simulation for AAM pilot workload studies*. NASA Aeronautics (NTRS). https://ntrs.nasa.gov/

OECD. (2023). *Tourism trends and policies: Greece 2023*. Organisation for Economic Co-operation and Development. https://www.oecd.org/cfe/tourism/oecd-tourism-trends-and-policies.htm

Volocopter GmbH. (2025). *Developing the regulatory framework for Urban Air Mobility—EASA's innovative EU approach*. https://www.volocopter.com/en/blog/developing-the-regulatory-framework-for-uam

5 Economics of AAM—eVTOL Networks

Dimitrios Ziakkas, Jam Hartley, and Amin Vafadar

5.1 INTRODUCTION

Advanced Air Mobility promises to shrink cities, not by erasing distance but by collapsing time. The economic question is whether that promise can be delivered reliably, affordably, and at scale for real travellers with ordinary constraints—budget, schedule, comfort, and predictability. This chapter examines AAM through that lived lens. Rather than treating eVTOLs as an abstract technology, we organize the economics around the journey: the price a passenger is willing to pay for minutes saved; the cost for an operator to produce those minutes at an acceptable risk; and the public's calculus when trading scarce urban space, grid capacity, and noise headroom for a new transport layer.

Our analysis proceeds from unit economics upward. We begin at the aircraft level—energy draw by phase of flight, the cost of electricity at the plug, battery life consumed per cycle, line-maintenance touch time, and the reliability penalties that attend any young fleet. We then widen to the node: vertiport capacity is not just pad geometry; it is an operations function of dwell times, charging power and state-of-charge windows, safety envelopes, and the choreography of people and vehicles. From there, we shift to the network, where utilization is earned or lost in dispatch planning, buffer strategy, and the way corridors, weather minima, and airspace services translate into predictable block times. At each tier, we carry both sides of the ledger—CAPEX and OPEX—because financing structure, depreciation policy, and battery replacement rules can change the same operation from marginal to viable.

Demand is treated with equal discipline. We distinguish curiosity from durable willingness to pay, and we confront the elasticity of price versus time saved across archetypal markets: airport connectors, medical and public-interest missions, premium commuter links, and tourism flows. Importantly, we translate qualitative acceptance into quantitative constraints: opening hours shaped by community dialogue, noise-aware routing that lengthens track miles, and environmental commitments that bind energy sourcing and reporting. The resulting picture is not a single "answer," but a map of viable operating envelopes—where seat-kilometer costs can be covered by fares and where externalities are priced and mitigated rather than ignored.

Finally, we anchor the discussion in trajectories, not snapshots. Early services will carry the weight of high capital costs, cautious duty cycles, and conservative schedules; learning effects, supply-chain maturation, and software-driven optimization

 DOI: 10.1201/9781003685555-5

are expected to move the curves. We therefore present economics as a path: from demonstration to dependable service, and from dependable service to density. The signposts along that path—utilization thresholds, turn-time limits, battery replacement cadence, energy and demand hedges, and financing covenants—are stated plainly so that public and private actors can make transparent choices about where to invest and how to scale.

5.2 REVENUE STREAMS AND PRICING MODELS FOR eVTOL OPERATIONS

Volodymyr Bilotkach

5.2.1 Business Cases for eVTOL

Generally speaking, the projected business cases for eVTOL segment of the aviation ecosystem include those replacing both "traditional" aviation vehicles as well as those infringing on surface transportation. The success of those businesses will obviously depend on the relevant cost, demand, and regulatory environment (Aldemir & Mavi, 2024). At the high level, eVTOL vehicles can be employed in the following businesses:

- **Urban air mobility (UAM)/air-taxi services:** Catering to higher-income commuters and business travelers, as well as to some extent to the general public, these services will provide short, on-demand flights within a metropolitan area. The following segments appear to be conducive to the development of businesses of this kind:
- Metropolitan areas suffering from road congestion.
- Metropolitan areas with a considerable share of high-income residents.
- Metropolitan areas with well-developed professional service industries (financial services, insurance, legal services, etc.).
- Densely populated areas with significant bodies of water and limited surface transportation options across them (e.g., travel between Singapore and nearby Indonesian islands, currently served by ferries).

Businesses of this kind will operate in parallel with their traditional surface transport competitors (taxis, ride-sharing companies, and ferries). The high cost of providing those services will likely make them affordable primarily to the higher-income travelers (Bridgelall, 2023).

- **Airport shuttle:** This business case would represent a reformatting of the helicopter services between vertiports located in central business districts (CBDs) and major airports. For instance, helicopter services have been provided between Manhattan vertiports and airports serving New York City area since as early as 1965. The combination of low demand and high cost of such flights meant that services were not provided consistently, with operators occasionally going out of business.

Such services can thrive in densely populated metropolitan areas with a significant number of high-value-added professional service businesses clustered within a CBD) and airports with surface transport accessibility issues (such as road connection, lack of rail connectivity with the CBD, and/or considerable distance between CBD and airport). It is therefore not surprising that New York City became one of the first locations where such services were commercially offered. We can fully expect eVTOL vehicles to replace airport shuttle helicopter services, where those are provided. Depending on the cost advantages of eVTOL services, as well as local noise and other regulatory restrictions, we can see eVTOL vehicles being used for such services in metropolitan areas where helicopter airport shuttle flights were never offered before. Airport shuttle services will serve to connect passengers to flights to/from the major airports. Therefore, we can fully expect such eVTOL flights to be provided in cooperation with the major airlines serving the airports. Moreover, the airlines themselves might decide to enter this segment of the business.

- **Regional air mobility:** Here, eVTOL vehicles can encroach on and/or expand the markets served by commuter airlines (Part 135 carriers in the United States). These are short-haul markets that are too thin to attract regular scheduled services by the regional jets or turboprop aircraft. Such services can earlier be stand-alone operations serving origin-destination traffic or provide passengers with opportunities to connect to regional or mainline jet-operated flights out of major airports.

Services on some of the markets in question are currently subsidized by public funds (through programs similar to Essential Air Services in the USA or Public Service Obligations in the European Union). Thus, eVTOL operators wishing to enter the Regional Air Mobility segment of the market may be able to bid for the right to operate those services. Regulatory requirements for the carriers to qualify for such operations may impose additional burdens on eVTOL operators, reducing their competitiveness vis-à-vis incumbent commuter carriers.

- **Emergency and medical services:** There is a good business case for eVTOL vehicles to be used for such applications as rapid response in urban or remote environments for medical evacuations, organ or blood transport, or disaster response. These vehicles can both replace helicopters where those are currently used and expand the size of the market, allowing for operations where such have previously been either commercially non-viable or unsafe for helicopters due to geography or the nature of the built environment.

Depending on the regulatory environment in specific jurisdictions, the private sector might be blocked from providing emergency and medical services. Thus, while such a business case can be identified, it might not exist in some countries.

- **Cargo and logistics operations:** The drone delivery concept has been considered for quite some time. One could say that logistics operations facilitating first-mile or last-mile delivery were perhaps the first business case considered and described for such vehicles. A variety of vehicles can be used in this business segment, from larger vehicles for transportation between airports and/or vertiports to small pilotless drones for delivery to the final customer.

Cargo operations using eVTOL vehicles will likely be limited to time-sensitive or high-value cargo, due to the likely cost differential between air and surface cargo transportation. We can expect higher utilization rates of cargo delivery vehicles in metropolitan areas prone to surface road congestion, as well as to/from islands with limited surface transportation options, other things being equal. We can fully expect cargo integrators (such as FedEx or UPS) to incorporate such vehicles into their value chains. In fact, these companies are currently actively studying such possibilities. The emergence of independent companies operating fleets of eVTOL vehicles for cargo operations cannot be ruled out either.

- **Tourism and scenic flights:** Helicopter tours are currently being offered in numerous popular tourist locations. Given likely cost savings to be associated with eVTOL vehicles, we can expect gradual replacement of helicopters in this market segment. According to some estimates, the size of the global helicopter tours market is projected to surpass $1 billion by 2030 from about $660 million in 2022. This market segment offers considerable opportunities for the eVTOL industry (Chinthoju, 2025).
- **Corporate and private mobility:** We do not expect corporate-owned eVTOL vehicles to replace many corporate jets. Yet, replacement of corporate helicopters with these vehicles is a clear possibility. Fractional ownership models, similar to those present in the business jet market, are likely to appear in this market segment as well.

Use of eVTOL vehicles for private recreational use and/or commuting represents another business opportunity for manufacturers of those vehicles. For a variety of reasons, mostly having to do with cost and practicality, private ownership of helicopters is not a very popular option among aviation enthusiasts. Rotorcraft currently represents about five percent of all general aviation aircraft registered in the USA. The new eVTOL aircraft have the potential to change this equation by reducing ownership and maintenance costs as compared to helicopters while offering the owners operational advantages of vertical take-off and landing capability. With over 200,000 general aviation aircraft registered in the United States alone, the potential market is huge.

5.2.2 Market Structure

On the topic of market structure, we need to discuss the potential for competition in the market for both eVTOL manufacturers and operators, the latter focusing on the various business cases we identified in the previous subsection (Ison, 2024).

As with other segments of the aircraft manufacturing industry, we can expect eVTOL production industry to be relatively concentrated, and more so the larger the vehicles in question are. Manufacturing of aircraft and their key components (such as engines, landing gear, etc.) is associated with considerable economies of scale—meaning that the firm has to be large to be efficient. We largely expect this to be true for the new eVTOL market segment. At the same time, similarly to what we currently observe in the markets for helicopters and general aviation aircraft, concentration in the eVTOL industry will be less than what we observe with the manufacturing of mainline and regional jets. In addition to global players, we can expect the emergence of smaller regional manufacturers, focusing on sales in geographic proximity to their manufacturing facilities. This mirrors what we currently observe in the helicopter and general aviation aircraft manufacturing industries.

While currently prospective eVTOL manufacturers represent independent start-up companies, we also see Airbus entering the game in this market sector. This manufacturer is already present in the helicopter segment and has been working on developing electric aircraft. It is thus quite natural for Airbus to also try to compete in the emerging eVTOL market segment. Boeing is participating in the industry through its wholly owned Wisk Aero subsidiary. General Electric—one of the leaders in the jet engine market segment—is also actively developing aircraft for eVTOL market. Going forward, we can expect other helicopter and general aviation aircraft manufacturers branching out into eVTOL. As I noted in the previous subsection, the private mobility segment has the potential to generate considerable market volume, especially if more customers start viewing eVTOL vehicles as substitutes for general aviation aircraft for recreational use.

Similar to what we usually observe in emerging industries, we can expect market concentration levels to decline as the industry matures. Currently, we have over a dozen companies vying for a spot in this future industry. While not all of them might be direct competitors with each other, depending on the types of vehicles they are developing and the market segments they are planning to serve, by the time the eVTOL industry opens in earnest, some of the companies currently on the scene will likely be gone through bankruptcy and/or consolidation. Further consolidation can be expected among the future eVTOL manufacturers as the industry matures. In addition to straightforward horizontal mergers among the current and future independent start-ups, we may observe general aviation and helicopter manufacturers getting into this industry by purchasing more successful start-ups (Goyal et al., 2021).

Aircraft leasing companies are projected to be active players in the eVTOL industry, especially at the earlier stages of the industry's development. Currently, eVTOL industry is subject to considerable risks—it is not known, for instance, which vehicles currently under development will be certified and when; the capital and operating costs are also subject to uncertainty; the regulatory environment has not been fully developed, etc. By including eVTOL vehicles in their portfolios and mixing them with traditional aircraft, aircraft lessors can absorb some of those risks. For instance, Avolon—one of the major players in the aircraft leasing industry—ordered 500 eVTOL vehicles from the manufacturer Vertical Aerospace in 2021. Valued at $2 billion—according to Avolon's website—this portfolio represents a fraction of the value of the lessor's fleet of over 1,000 mainline aircraft (with nearly 600 in ownership and over 400 representing orders or options).

As eVTOL aircraft will come at a fraction of the cost of mainline jet aircraft, as the industry matures, and as operators obtain a better understanding of the risks involved, we can expect lessors playing a less significant role over time. Nevertheless, leasing companies will not exit this market segment altogether. In fact, while exact data are not available, we have over the last decade been observing an increasing role of lessors across all the market segments, including general and business aviation.

When thinking about market structure in eVTOL operator market segments, we can draw parallels with similar industries in both aviation and surface transportation ecosystems. Let us go over the relevant business cases identified earlier (Straubinger et al., 2021). We will ignore the private mobility business case, as we can expect most vehicles in this category will be privately owned. Yet, we acknowledge that fractional ownership businesses might appear in this segment as well.

- *UAM* and/or *Air-Taxi* services are somewhat similar in nature to shared-ride enterprises. Yet, unlike Uber/Lyft/Grab or similar companies, which are essentially IT enterprises operating platforms, with drivers joining those platforms as independent service providers, eVTOL vehicle cost will likely lead to the establishment of vertically integrated enterprises. It is hard to predict whether we can expect the emergence of mega-platforms similar to Uber or Lyft in this market, or if air-taxi services end up being provided by multiple smaller regional platforms. Given that we expect those platforms to be vertically integrated, economies of scale will likely mean that outside of the largest metropolitan areas, we can expect a single platform to dominate. Another interesting open question is whether local governments may decide to impose economic regulation on the air-taxi segment, similar to what we regularly observe in the surface taxi markets. I do not expect this to happen, as air-taxi clientele will represent a small share of the voters. Regulators will thus have little incentive to impose economic regulation on this business.
- *Airport Shuttle* services have previously been provided by helicopter companies in a limited number of large metropolitan areas characterized by surface road congestion. We expect the emergence of eVTOL vehicles to increase the number of locations where such services become economically viable. In a limited number of locations, such as New York City or Los Angeles, the market size might end up being sufficiently large to accommodate two or more competitors; yet, in most cities, we anticipate a single operator providing such flights. At airports that serve as hubs for major network airlines, we can expect hub operators incorporating eVTOL airport shuttle services into their networks, including assigning their flight numbers, allowing passengers to book those flights through their websites, and incorporating them into network revenue management systems. Hub operators might go as far as to market those shuttle flights and even entering into long-term contracts with the operators, similar to those they have with the regional airlines in the USA. Network airlines acquiring ownership stakes or purchasing airport shuttle service eVTOL operators also cannot be ruled out in the longer term.
- *Regional Air Mobility* markets are served by both commuter and regional airlines. It appears that, taking the US case as an example, eVTOL vehicles will allow operators to provide services under the less stringent Part

135 Federal Aviation Regulations, as compared to Part 121 Regulations that apply to regional airlines operating larger aircraft. We can foresee the following trajectories for the development of this market segment. First, commuter carriers may either expand their fleet with eVTOL vehicles or gradually replace their existing fleet of jet and turboprop aircraft with eVTOLs. Second, new operators can emerge, who will compete with the incumbent commuter carriers using their fleet of eVTOL vehicles. Third, both of the two scenarios can play out simultaneously. We should note that a major factor determining the extent to which eVTOL vehicles end up playing a role in Regional Air Mobility market segment will be the development of electric or hybrid aircraft. Small aircraft would be the first logical target for deployment of this technology. While eVTOL vehicles are expected to attract lower maintenance costs as compared to helicopters, small electric aircraft are expected to provide an operating cost advantage over eVTOL. Further, at least initially, the electric aircraft purchase price is likely to be lower as well—we are talking about essentially an established platform with a different engine as compared to a largely new platform. Thus, there is a considerable chance that developments in the aircraft technology might limit the viability of eVTOL vehicles to routes requiring their vertical take-off and/or landing capability.

- The *Emergency and Medical Services* market segment is a tricky one, as we indicated earlier. In some jurisdictions, those services will be provided by the government. Yet, even in countries with single-payer healthcare systems (that single payer being government or government-regulated entity), private companies can contract for the provision of medical services such as airlifting patients to the hospitals. Market structure in this segment will be largely driven by a combination of the projected market size and the regulatory environment.
- In the *Cargo and Logistics Operations* market segment, eVTOL vehicles will most likely be incorporated into the existing cargo integrators' business structure. These vehicles are most viable for the express delivery segment, where time is of the essence. They will largely facilitate first- and last-mile connectivity. The express delivery market is largely dominated by the integrator companies (such as FedEx, UPS, and DHL), which require scale for efficiency and the ability to serve a wide variety of markets. We can also expect the emergence of smaller-scale regional delivery companies, which will contract with the large integrators to provide first- and last-mile connections in smaller and/or remote markets. We currently observe such arrangements in some regions. Simply put, in smaller markets, it may be more cost-effective for each of the integrators to contract with the same local delivery company for first- and last-mile connectivity instead of relying on their own vehicles.
- *Tourism and Scenic Flights* markets currently served by helicopters will likely see many of those vehicles replaced by eVTOL craft. Depending on the market size and operating costs of eVTOL vehicles, we might see a more competitive environment on some of those markets.

5.2.3 Pricing Models for eVTOL Operators

At the basic level, services can be priced *linearly* or *nonlinearly*. Linear pricing is the term used by economists to denote the simple case of pricing each unit of goods or service equally for all customers. The service providers can also engage in *price discrimination*, defined as applying different prices to different customers of the same product or service, with these price differences not being justified by the differences in the cost of providing those services. An equivalent definition of price discrimination is setting prices in such a way as to obtain different markups from different customers. Note that with the latter way of looking at this phenomenon, we can see price discrimination even where prices are the same: if costs of providing the service to different customers differ, markups will be different too.

Nonlinear pricing refers to different methods of setting the prices in such a way that the per-unit price varies with the quantity purchased. The eVTOL business holds considerable potential for the application of those techniques. Let us review the commonly used nonlinear pricing tools in the context of eVTOL business.

- **Quantity discounts/block pricing:** This is a rather ubiquitous practice of offering a lower per-unit price with larger orders. An example of the use of this tool may include a commuter buying a bundle of 20 rides per month at a lower average fare than buying single trips. Such discounts could be applied to most business use cases we have described earlier.
- **Two-part tariffs:** A two-part tariff involves charging a fixed fee for access to the service, followed by a fixed per-unit price. Most utilities use versions of this pricing tool. An example in the eVTOL business can include a passenger paying a monthly subscription fee followed by a reduced per-ride fare. Occasional riders may stick to single fares, while frequent users will find value in the subscription. Cargo capacity can be priced in a similar way.

Successful application of two-part tariffs requires a considerable level of understanding of the customer demand. The issue here is to design the two-part tariff in such a way that the regular commuter would prefer the subscription as described in the previous paragraph to paying regular single fares. Note also that block pricing is similar to two-part tariff—essentially, block pricing involves a fixed fee that gives the customer access to a certain quantity of product or service at zero price. Two-part tariffs are generally a more flexible pricing tool than block pricing.

An extension of two-part tariffs is a multipart tariff, which is defined as a nonlinear tariff with increasing block prices. This pricing technique involves a fixed access fee. However, the unit price under the multipart tariff rises after a certain consumption threshold is reached.

- **Bundling or package pricing:** This technique involves multiple products or services being sold together, usually at a discount. Holiday packages that include a flight, a hotel room, and a car rental would be a typical example of this pricing technique. Another commonly used example of bundling is the meal deals at fast-food restaurants. An eVTOL operator could apply package pricing by partnering with ground transport. A "home-to-office package" would then include a car pickup, eVTOL flight, and last-mile shuttle at a combined price that is lower than buying each leg separately.

In economics, we distinguish between two types of bundling. *Pure bundling* involves services that are only available as a package, with individual components of the package not being sold separately. This type of bundling was common in the airline industry until the airlines started unbundling their products and selling add-on services. At the same time, we should note that some of the add-on services are not available for purchase individually—you cannot purchase seat selection without paying for a seat on the flight. *Mixed bundling* involves services being available for purchase both as a package and individually. In the above "home-to-office package" example, for instance, the service provider will likely use a mixed bundling approach—customers who have access to a personal or company car might still choose to only pay for the flight.

- **Versioning/quality-differentiated pricing:** This technique involves offering different "tiers" of service at different per-unit prices. An airline offering economy, premium economy, and business class cabins would be a simple example of such a pricing approach. This is one strategy that eVTOL operators could adopt from the airline industry.

In principle, most of the revenue management techniques used by the airline industry also fall within this category. When the same seat is priced differently, in many cases, such different prices are accompanied by different refundability conditions, meaning that different fare categories represent different "tiers" of service. We can definitely expect eVTOL operators to adopt revenue management for both passenger and cargo capacity.

- **Loyalty and membership programs:** Frequent flier programs (FFPs) are widespread in the airline industry. Whether eVTOL operators choose to develop their own programs will depend on their scale of operations. Services linked with the airline operations, such as airport shuttle, will likely partner with the airlines' FFPs or maybe even utilize the airlines' customer loyalty programs instead of creating their own.

At the earlier stages of FFPs, the number of credits the customers accumulated was proportional to the flight distance. Currently, most such programs award credits in proportion to the amount of money spent by the customer. For eVTOL businesses, FFPs would most likely award credits based on either the number of trips or the amount of money spent.

- **Peak-load pricing:** Peak-load pricing is a technique usually used in industries featuring fixed capacity and seasonal demand. At the basic level, this technique involves charging higher prices in the high-demand season and lower prices in the low-demand season. In the context of revenue management of sparse seat or cargo capacity, peak-load pricing will involve not making lower-priced fare categories available during the high-demand season. Surge pricing is another technique that can be used for air-taxi services, similar to how this tool is utilized by ride-share companies.

5.3 CAPITAL EXPENDITURE (CAPEX) AND OPERATIONAL EXPENDITURE (OPEX)

Volodymyr Bilotkach

In this section, we will outline the key categories of capital and operating expenditures that eVTOL businesses are likely to face (Wu & Zhang, 2021). Capital expenditure (CAPEX) represents upfront long-term investments into capital goods. Those capital goods are, in most cases, eligible for amortization over time after purchase. Some of the capital goods can be leased rather than purchased. We can in particular expect the leasing market for eVTOL vehicles to develop, with both current aircraft lessors and new companies specializing in the leasing of eVTOL vehicles playing a role there. Whether the original equipment manufacturers (OEMs) of eVTOL vehicles may also play the role of lessors is currently anyone's guess. For context, while aircraft OEMs (such as Boeing, Airbus, and Embraer) are not involved in the leasing business, the engine OEMs (General Electric, Rolls-Royce, etc.) act as engine lessors.

- **Capital expenditure:** We can identify the following categories of capital expenditure in eVTOL business:
- *Aircraft development* expenditure is currently being borne by several companies working to bring the eVTOL vehicles to the market. These expenditures include research and development (R&D) and certification costs.
- *Aircraft acquisition* will be a key expenditure for eVTOL businesses. We should note that this category can include either outlays for aircraft purchases or lease payments. In the airline business, for instance, lease payments are usually included in the aircraft ownership cost category. We should note that a relatively smaller role of the "fuel" cost component at eVTOL businesses will imply that the aircraft acquisition cost will be a relatively more important cost component than aircraft ownership currently is for the airlines.
- *Vertiports and ground infrastructure* costs might not be borne directly by the eVTOL businesses. This infrastructure can be constructed using public funds, private operators' money, or public-private partnerships (PPPs).
- *Construction or adaptation of take-off/landing sites, passenger terminals, and hangars* is another capital expenditure item that may not necessarily be borne by the eVTOL operators. Nevertheless, since eVTOL-related infrastructure projects will in most cases be smaller-scale endeavors as compared to the airports, we can expect some larger-scale operators to also bear the cost of related infrastructure.

- *Charging infrastructure* includes such items as *fast chargers, grid upgrades,* and *energy storage.* For grid updates and energy storage, eVTOL businesses may be able to piggyback on the relevant infrastructure investments associated with the spread of electric vehicles. Thus, only fast charging infrastructure might end up being the expenditure item specific to eVTOL. Even here, similarities between electric and eVTOL vehicle charging could help reduce the related costs.
- Construction or upgrading of *maintenance and operations bases* is an expense item that could be borne by either eVTOL operators or specialized maintenance, repair, and overhaul (MRO) companies. In either case, the operators will have to decide between conducting maintenance in-house versus outsourcing the service. As with the airlines, larger operators will be more likely to do in-house maintenance, which will necessitate expenditure on the maintenance bases.
- Spending on *workshops, tooling, and diagnostic equipment* is a common expenditure item in the aircraft operators' business, as it spending money on maintaining the *spare parts inventory.*
- *IT systems* are also a common item of CAPEX expenditure in aviation. While some businesses might find it practical to develop those systems in-house, in most cases, expenditure on IT systems will involve purchasing and maintaining either off-the-shelf or customized solutions.
- For service-oriented eVTOL businesses, such as air taxis, development and maintenance of *booking apps and fleet management software*, as well as the *integration with ride-hailing platforms*, will be a CAPEX item.
- Last but definitely not least, eVTOL businesses will require *specialized support equipment*, such as *simulation and training equipment* and *ground handling equipment.*
- **Operating expenditure (OPEX):** Operating expenditures an eVTOL business can expect to face will be similar to those currently faced by the aircraft operators. At the same time, some clear differences will be present. We can define the following key OPEX categories here:
- *Cost of charging eVTOLs* will be equivalent to the fuel bill faced by the aircraft operators of the present time. It would be a stretch to suppose that the charging cost will constitute up to a quarter of the operator's expenditures, similar to the relative size of the fuel bill currently faced by some airlines. At the same time, charging cost can potentially be variable, with the operator facing peak/off-peak grid rates, as well as higher rates for fast charging.
- *Labor cost* is another key OPEX item for the airlines. For eVTOL businesses, this cost item will also be very important. The components of labor cost will include pilot salaries (until fully autonomous vehicles become available), ground crew, dispatchers, vertiport staff, and customer service personnel. Labor cost includes salaries or wages, benefits, allowances, and any payroll taxes that are charged in many jurisdictions.
- Related to the labor cost described earlier is the *training cost.* This includes ab initio (where necessary) and ongoing pilot, engineer, and ground staff training and certification.

- *Maintenance* expenses will include such items as routine inspections, battery replacements, and airframe and avionics upkeep. In the near future, battery replacements will likely be the costliest maintenance item. As with most novel technologies, we can expect the battery replacement cost to decrease over time.
- *Financing costs* include interest payments on any debt instruments the business may issue. This usually includes aircraft mortgages, loans (secured and unsecured), and bonds. At the early stages of development, eVTOL businesses can expect to face somewhat elevated borrowing costs, as the financial sector will have a low degree of familiarity with this business. As eVTOL sector develops and the financial sector becomes more accustomed to dealing with the companies operating in this business, the borrowing cost can be expected to fall.
- *Insurance* will be a significant cost in the early stages of eVTOL business development. Operators will need coverage similar to that of airlines—hull, liability, and passenger policies—but liability insurance will differ in important ways. Because eVTOLs will frequently operate in urban environments, the risks of damage to property and infrastructure, as well as injuries to people on the ground, will be considerably higher than in traditional airline or even helicopter operations.
- Charges for using aviation infrastructure will include *airport/vertiport charges* and *air navigation fees*. We should note that accommodating eVTOL vehicles into the aviation ecosystem might require a different structure of the aviation infrastructure charges as compared to what airlines currently face. Airport charges typically include a fixed component, tied to the aircraft's maximum take-off weight (MTOW) in less than a direct proportion, as well as the variable component, or per-passenger charges. En route air navigation charges include weight (also linked to the MTOW in less than a direct proportion) and distance component, whereas terminal navigation fees are linked to the MTOW only. In some countries—most notably, the USA—the cost of providing air navigation services is covered through passenger taxes. We can expect that eVTOL vehicles will likely face fixed infrastructure charges, differentiated by vehicle class, which will likely be linked to their weight.
- *Vertiport operation* costs will include such items as rent, utilities, security, cleaning, and ongoing facility maintenance.
- Many eVTOL businesses will face *marketing and customer acquisition costs*. These include such items as branding, partnerships with mobility platforms, and the cost associated with the loyalty programs.
- Last but not least, *administrative overhead* includes such costs as licensing, IT support, management, and back-office operations.

In summary, the cost structure of eVTOL businesses will combine substantial upfront capital investment with recurring operating expenditures that mirror, but also diverge from, those of conventional aviation. On the capital side, the most significant outlays will be for aircraft acquisition—whether through purchase or leasing—supported by investments in vertiports, charging networks, IT platforms, and specialized equipment. Some of these costs may be shared with public authorities or infrastructure partners, but operators will nevertheless need to plan for sizeable commitments, particularly in aircraft ownership and technology systems.

On the operating side, costs will be shaped by familiar categories such as energy, labor, maintenance, financing, and insurance, but with nuances unique to eVTOL operations. Electricity will replace fuel as the central energy expense, while battery replacement is expected to dominate maintenance costs, at least in the near term. Insurance premiums are likely to be elevated given the risks of urban operations, and financing costs may initially be higher due to industry novelty. Over time, however, as the sector matures, borrowing costs, maintenance expenses, and insurance rates can be expected to decline, improving the overall cost profile and sustainability of eVTOL operations.

5.4 ROI METRICS AND ECONOMIC FORECASTING FOR AAM PROJECTS

Volodymyr Bilotkach

When evaluating Advanced Air Mobility (AAM) projects or businesses (eVTOL operators, vertiports, infrastructure, or integrated ecosystems), traditional ROI metrics can be used, but they often need tailoring to reflect long lead times, regulatory risks, and uncertain demand (Franciscone & Fernandes, 2023). Forecasting for eVTOL projects and businesses, especially at the earlier stages of the industry's development, will be associated with considerable uncertainty. We advocate for extensive use of scenario forecasting here.

- **Financial ROI metrics:** Traditional tools of cost-benefit analysis and financial analysis are well suited for the evaluation of eVTOL businesses and projects. Yet, these techniques must be adapted to account for the unique characteristics of this emerging sector. Unlike established industries with predictable demand, stable cost structures, and mature regulatory frameworks, eVTOL ventures face high upfront capital expenditures, evolving operating costs, and significant uncertainty around the speed of market adoption. Investors, financiers, and policymakers will therefore look to a range of return on investment (ROI) metrics not only to assess the profitability of these projects but also to compare them with other infrastructure and mobility opportunities.

The most common tool of the cost-benefit analysis is the *Net Present Value (NPV).* NPV captures the present value of expected cash inflows minus outflows over the lifetime of an investment, discounted to reflect the time value of money. For eVTOL projects, cash outflows are concentrated in the early years due to aircraft acquisition, vertiport construction, charging infrastructure, certification, and regulatory compliance. Cash inflows, by contrast, depend on uncertain passenger or cargo demand, utilization rates, and the trajectory of regulatory approvals. A positive NPV indicates that the project generates value above its cost of capital. The result of the NPV modeling depends on the values of various variables: load factors, pricing strategies, maintenance and other operating costs, among others. Sensitivity analyses around these variables will be essential, since small shifts in assumptions about passenger uptake or battery replacement cycles can significantly affect the outcome.

Closely related to NPV is the *Internal Rate of Return (IRR),* which represents the discount rate at which the NPV of an investment equals zero. The IRR is particularly useful in comparing AAM investments with other infrastructure or mobility projects competing for capital. For example, investors may weigh the IRR of an eVTOL operator against that of an airport terminal expansion, a high-speed rail project, or even a ride-hailing platform. Because eVTOL businesses combine elements of aviation and urban transport, their IRR will be benchmarked not only against aviation returns but also against those of urban mobility and technology-driven ventures. The challenge is that early-stage eVTOL projects may show modest or even negative IRRs due to long certification timelines and gradual market ramp-up, which may make them less attractive unless investors are willing to accept higher risk in exchange for long-term upside.

The *Benefit-Cost Ratio (BCR)* is another traditional tool. Unlike NPV or IRR, which focus on absolute financial returns, the BCR expresses the relationship between the present value of benefits and the present value of costs as a ratio. A BCR greater than one indicates that benefits exceed costs, while a value below 1 suggests the opposite. This metric is especially useful in the eVTOL context because it enables comparisons between projects of very different scales. For example, a small vertiport serving a single urban corridor and a larger regional eVTOL hub may both show positive NPVs, but their BCRs will reveal which project delivers more benefit per dollar invested. By normalizing outcomes across project sizes, the BCR provides a common yardstick that is particularly relevant when public funding or PPPs are involved.

More risk-averse investors or those with shorter planning horizons might be interested in the *Payback Period.* This metric calculates the time required for cumulative discounted cash inflows to equal the initial investment. The eVTOL businesses may feature longer payback periods compared to more conventional mobility projects, as commercialization cycles in aerospace are often protracted and heavily influenced by regulation. The payback period is a particularly relevant metric for venture capital and private equity investors who need to evaluate whether returns will materialize within their investment horizons. Shorter payback periods are generally more attractive. However, in the context of eVTOL business, investors may need to reconcile themselves with longer horizons given the transformative potential of the technology.

A more comprehensive and rather common measure of capital efficiency is *Return on Invested Capital (ROIC).* This ratio, calculated as operating profit after tax divided by invested capital, demonstrates how effectively a company uses its capital base to generate returns. A strong ROIC indicates that the company is generating sufficient returns relative to the capital employed. This is an important consideration for an industry that will compete not only with other transport modes but also with such capital-intensive sectors as renewable energy or telecommunications.

EBITDA Margin and Cash Flow Margin become critical indicators of financial sustainability once the business starts operating. EBITDA margin measures earnings before interest, taxes, depreciation, and amortization relative to revenues. Cash flow margin assesses how effectively revenues translate into operational cash flows. For eVTOL businesses, these metrics highlight the ability of operators to cover financing costs and generate operating surpluses. In the early years, margins may be compressed by high insurance premiums, training costs, and financing expenses. Over time, however, these pressures may ease as the industry matures and scale effects are realized.

- **Operational ROI metrics:** The operational metrics of eVTOL businesses mirror, to some extent, those commonly used in the airline industry. These measures reflect the interplay between technology, infrastructure, and market demand and provide a practical lens for evaluating the sustainability of business models in this emerging sector.

Cost per Available Seat-Kilometer (CASK) reflects the average operating cost incurred for transporting a seat—whether occupied or not—one kilometer. In the USA, CASM rather than CASK is used, reflecting cost per available seat mile. In commercial aviation, CASK is the commonly used average cost benchmark, which can be used by eVTOL business. The equivalent air cargo metric is the Cost per Available Ton-Kilometer. In commercial aviation, a standardized cargo density of 161 kg per cubic meter is used to translate available cargo volume into available cargo capacity by weight.

The relevant revenue metric is the *Revenue per Available Seat-Kilometer (RASK)*, which measures the average revenue earned per unit of seat-distance flown. When RASK is higher than CASK, the aviation business shows profitability. The business loses money if the reverse is true (CASK is larger than RASK).

Another critical metric associated with the viability of an aviation business is the *Load Factor*, which indicates the proportion of available seat or cargo capacity that is actually filled. In the eVTOL market, achieving high load factors will depend on sophisticated demand forecasting, integration with digital booking platforms, and effective scheduling of both peak and off-peak flights. Low utilization rates not only depress revenues but also lengthen payback periods, potentially making the entire business model unviable. Conversely, high utilization maximizes revenue generation and spreads fixed costs over a larger base, improving overall returns.

Finally, *Fleet Utilization* assesses revenue generation at the level of each individual aircraft relative to its capital and operating costs. For an eVTOL operator, this metric underscores the importance of maximizing flying hours per vehicle through high-frequency scheduling and minimizing downtime due to maintenance or charging. This metric is widely used in the airline industry. An airline typically measures fleet utilization by aircraft type. Strong fleet utilization not only improves profitability but also provides a more attractive risk-return profile for lessors, financiers, and investors considering exposure to the sector.

- **Strategic ROI metrics:** *Strategic ROI metrics* capture broader economic, social, and ecosystem-level returns that extend beyond the financial statements of individual operators. They are particularly relevant because the eVTOL sector sits at the intersection of aviation, urban mobility, and sustainability, where the value proposition extends to governments, communities, and partner industries as much as to investors and shareholders.

One such measure is the *Economic Impact Multiplier*, which reflects the broader contribution of eVTOL deployment to GDP, employment, and trade. Governments often rely on this type of metric when evaluating infrastructure investments such as vertiports or grid upgrades. For instance, new eVTOL services may create direct employment in operations, maintenance, and customer service, while also generating indirect jobs in supply chains and induced effects through local spending. Catalytic benefits, including attracting new business investment and enhancing tourism, further extend the economic impact. This multiplier perspective is critical in policy discussions, as it positions eVTOL not merely as a transport option but as a driver of regional economic competitiveness.

The *Network Effect* captures the incremental value created as eVTOL route networks expand. A single vertiport connection may provide convenience, but the true value of eVTOL emerges when multiple routes create a web of interconnected mobility options. Each additional route increases the utility of the network for both passengers and operators, reinforcing demand and improving fleet utilization. This effect is particularly powerful for platform-based operators that integrate eVTOL services into broader multimodal ecosystems.

- **Forecasting of eVTOL projects:** Economic forecasting for eVTOL projects presents unique challenges that distinguish this emerging sector from more mature industries. Forecasting in aviation traditionally relies on decades of historical data, well-established demand patterns, and predictable cost structures. In contrast, eVTOL businesses operate in a context defined by regulatory uncertainty, evolving technology, and highly variable adoption rates. The absence of historical benchmarks means that forecasting must be approached with caution, and investors, policymakers, and operators alike need to recognize the limitations of point estimates. Instead of relying on single deterministic projections, economic forecasts for eVTOL projects should employ scenario analysis to capture the wide range of potential outcomes that may unfold as the industry matures.

One of the key uncertainties in eVTOL forecasting lies in demand estimation. While there is broad recognition of the potential time savings and convenience offered by urban air mobility, it is difficult to predict how quickly consumers will adopt these services and whether eVTOLs will be positioned as premium products or evolve into mass-market solutions. Will early adopters remain limited to high-income business travelers, or will costs fall sufficiently to compete with ride-hailing and short-haul ground transport? Similarly, willingness to pay will vary widely across regions and customer segments. Forecasting demand thus requires multiple scenarios: a high-demand case with rapid adoption, a baseline case with gradual uptake, and a low-demand case where adoption remains niche for longer than anticipated.

On the cost side, operating expenses such as energy, maintenance, insurance, and labor remain highly uncertain. Battery replacement costs represent a significant unknown, as they will depend on the pace of technological improvement, the durability of early eVTOL models, and the evolution of the electric supply chain. Insurance premiums are also expected to be high in the early stages due to perceived risk but may decline as the industry establishes a safety track record. Scenario analysis allows forecasts to account for these cost variations, for instance, modeling cases where battery costs decline rapidly versus cases where they remain persistently high.

Regulation and infrastructure development introduce further uncertainty. Certification timelines are highly uncertain, and delays could postpone revenue generation by several years, sharply reducing the attractiveness of a project. Similarly, the pace at which vertiports, charging infrastructure, and digital air traffic management systems are deployed will directly influence the scalability of operations. Scenario analysis can explicitly incorporate these regulatory and infrastructure milestones, modeling a best-case scenario of early certification and rapid infrastructure rollout alongside more conservative, slow-growth cases.

Thus, economic forecasting for eVTOL projects requires moving beyond static projections to embrace scenario analysis as a central tool. The nascent nature of the industry, combined with technological, regulatory, and behavioral uncertainties, makes forecasting increasingly difficult. By constructing a range of plausible futures—varying demand adoption, cost trajectories, certification timing, and infrastructure rollout—stakeholders can better understand risks, identify resilience strategies, and make more informed investment decisions. Scenario analysis thus provides not only a more realistic view of the economic potential of eVTOL projects but also a framework for navigating uncertainty in this transformative sector.

5.5 CONCLUSION

Dimitrios Ziakkas, Jam Hartley, and Khalid Waseem

AAM networks scale when infrastructure, policy, operations, and demand are co-designed and phased against evidence. The practical blueprint that emerges across geographies and governance models is consistent: start small at high-value nodes; codify community-fit through zoning; invest early in grid/charging with storage; operate on structured routes tied into U-space/UTM; mix passenger and cargo to lift utilization; and expand hours, weather minima, and throughput as safety cases mature. Regulatory sandboxes, SORA-based authorizations, and government-orchestrated demonstrations shorten the distance between promising trials and permanent rules, especially in emerging markets. Smart-city integration—digital plus physical—turns eVTOL from a novelty into a dependable layer in the mobility stack.

Finally, scaling choices should be read through a business and benefits lens: price architecture, utilization, CASK/CASM versus RASK, payback, and scenario-based ROI guide where (and how fast) to build; societal metrics—jobs, access, reliability, emissions, and resilience—sustain public and political support. Cities and operators that instrument operations and publish performance will earn the privileges that allow density to rise—safely, credibly, and with visible value delivered to communities.

REFERENCES

Aldemir, H. Ö., & Mavi, H. (2024). UAM passenger business model development: The effect of aviation authorities' strategies and technological advancements. *Transportation Research Procedia*, *81*, 35–43.

Bridgelall, R. (2023). Forecasting market opportunities for urban and regional air mobility. *Technological Forecasting and Social Change*, *196*, 122835. https://doi.org/10.1016/j.techfore.2023.122835

Chinthoju, P. K. (2025). Cost per flight analysis of tilt-wing eVTOLs for urban mobility. *International Journal of Applied Engineering and Technology*, *6*(4), 110–120.

Franciscone, B. G., & Fernandes, E. (2023). Challenges to the operational safety and security of eVTOL aircraft in metropolitan regions: A literature review. *Journal of Airline Operations and Aviation Management*, *2*(1), 45–56.

Goyal, R., Reiche, C., Fernando, C., & Cohen, A. (2021). Advanced air mobility: Demand analysis and market potential of the airport shuttle and air taxi markets. *Sustainability*, *13*(13), 7421. https://doi.org/10.3390/su13137421

Ison, D. C. (2024). Taking flight or taking a pass? Exploring factors influencing consumer willingness to pay for eVTOL travel. *International Journal of Aviation, Aeronautics, and Aerospace*, *11*(1), Article 5. No DOI found.

Straubinger, A., Michelmann, J., & Biehle, T. (2021). Business model options for passenger urban air mobility. *CEAS Aeronautical Journal*, *12*(2), 361–380. https://doi.org/10.1007/s13272-021-00514-w

Wu, Z., & Zhang, Y. (2021). Integrated network design and demand forecast for on-demand urban air mobility. *Engineering*, *7*(4), 473–487. https://doi.org/10.1016/j.eng.2020.11.007

6 Integrating AAM into Global Aviation Ecosystems

Training, Operational, and Human Factors Perspectives

Gustavo A. Sanchez and Dimitrios Ziakkas

6.1 INTRODUCTION

The Advanced Air Mobility (AAM) concept of operations represents a major transformation to the current air transportation system. Earlier revolutions in aviation, such as the introduction of jet engines, radars, and glass cockpits, transformed air transport by introducing a single technology. In contrast, the AAM vision requires advances in multiple technological areas to materialize its benefits and bring to life a new era in aviation. The Electric Vertical Takeoff and Landing (eVTOL) vehicles play a central role in the AAM revolution, providing a new way to transport goods, passengers, and connecting essential services in regions, metropolitan areas, and underserved territories. The importance of AAM extends beyond new aircraft designs; it represents a shift toward integrated mobility solutions (NASEM, 2024). Civil aviation authorities worldwide stress that infrastructure readiness and certification pathways must advance in tandem with vehicle technology. AAM has the potential to reduce congestion in metropolitan areas, improve access to underserved communities, and enable faster emergency and medical response operations (NASEM, 2022). However, realizing these benefits depends on coordinated advances in propulsion, autonomy, digital air traffic systems, and regulatory harmonization. This systemic view positions AAM as not only a technological innovation but also a policy and societal milestone.

The eVTOL vehicles use advanced propulsion systems such as Distributed Electric Propulsion (DEP) and electric engines as primary power sources while enhancing aircraft control (Ugwueze et al., 2023). The flexibility provided for DEP during the aircraft design, engine reliability, and reduced noise footprint are among the major operational safety strengths of eVTOLs' use under the AAM concept of operations. Fully electric eVTOL vehicles require approximately 300–600 kilowatts during a conventional mission profile (NASEM, 2024). While hybrid configurations use conventional fossil fuels to enhance the aircraft's capabilities. Regarding the type of

DOI: 10.1201/9781003685555-6

energy used, eVTOL can be categorized as *Electric/Batteries, Electric Hybrid*, and *Electric/Hydrogen* (NASEM, 2024). The most common configurations for eVTOL vehicles include (1) Multi-rotor aircraft, (2) Lift + Cruise Aircraft, (3) Vectored Thrust Aircraft, and (4) other types of eVTOL (Su et al., 2024). A Multi-rotor eVTOL uses multiple rotors to generate lift. Instead of relying on wings for lifting purposes, this vehicle uses electrically driven propellers to control the aircraft throughout all phases of flight (Malpica et al., 2023). The *Lift + Cruise* category is a hybrid configuration that uses rotors for hovering and wings to generate lift during forward flight (Liu et al., 2023). Another eVTOL category is represented by *Vectored Thrust* vehicles. These vehicles use the same propulsion system for hovering and cruise, employing either tilt-rotors or tilt-wings during all phases of flight (Enenakpogbe et al., 2024). Other configurations include *Electric Helicopters, Hoverbikes*, and *Personal Flying Devices* (NASEM, 2024).

Unlike previous revolutions in aviation, the AAM concept of operations requires the convergence of technological breakthroughs in autonomy, energy management, Artificial Intelligence (AI), and airspace management. This chapter focuses on four technological domains that collectively may facilitate the integration of eVTOLs into the current air transport ecosystem. These include autonomy and AI-driven flight control systems, predictive maintenance and AI diagnostics, energy systems such as advanced batteries and charging infrastructure, and AI-powered airspace management through Unmanned Traffic Management (UTM). While each technological breakthrough has its own challenges and opportunities, their combined maturity and implementation are required for a safe and scalable revolution in air transport.

6.2 AUTONOMY AND AI-DRIVEN FLIGHT CONTROL SYSTEMS

Gustavo A. Sanchez

Early in aviation history, engineers and Original Equipment Manufacturers (OEMs) conceived the idea of equipping aircraft with autopilot capabilities. The Sperry Company introduced the gyroscopic stabilizer early in the 1910s (David, 2024). The Sperry Company's gyroscope was connected to ailerons and rudders to control heading and altitude without manual pilot input on the flight controls. As aircraft systems became increasingly complex, the aviation industry found in automation a tool to reduce pilot workload and improve the safety of air transport operations. During the 1960s, autopilots and instrument landing systems allowed automatic landings under low-visibility conditions, setting new benchmarks for automation reliability. Late in the 20th century, the introduction of fly-by-wire technology marked another revolution. The fly-by-wire technology transformed aircraft control by replacing traditional mechanical cables with electronic systems connected to servos (Abzug & Larrabee, 2002). In this design, pilot control inputs are converted into electrical signals, which are transmitted to actuators or servos that move the aircraft's control surfaces.

The Boeing 777 and Airbus A320 are well-known commercial aircraft that incorporate fly-by-wire technology to increase redundancy on control systems and improve computer-assisted stability during autopilot operations. The revolution brought about by fly-by-wire control systems in aviation was accompanied by the introduction of glass cockpits, where traditional analog indicators were replaced by digital displays and Flight Management Systems (FMS). This revolution enhanced aircrews' performance by improving situational awareness, reducing workload, and facilitating greater precision during the execution of complex navigation procedures, such as Instrument Landing System Category III approaches. Despite the benefits provided by fly-by-wire, glass cockpits, and FMSs, pilots operating aircraft equipped with these systems experienced steep learning curves while transitioning from a "hands-on" to a "System Management" approach for aircraft control.

In contemporary aircraft design and operations, highly automated cockpits continue the trend of shifting operators to system managers rather than manual controllers. State-of-the-art commercial aircraft, such as the Boeing 777X, incorporate a flight deck for aircraft control combined with advanced FMS, Controller-Pilot Data Link Communications (CPDLC), and provisions for future advanced Air Traffic Management (ATM) systems (Boeing, 2023). The AAM and eVTOL vehicles represent the next phase in the evolution of air transport (Goldfeld et al., 2024).

> *AAM is a transportation system that moves people and property by air between two points in the United States (U.S.) using aircraft with advanced technologies, including electric aircraft, or electric vertical takeoff and landing (eVTOL) aircraft, in both controlled and uncontrolled airspace.*
>
> **(FAA, 2023a)**

eVTOL vehicles are being developed to transport passengers, goods, and services in congested airspace during high-frequency operations (Semanjski et al., 2023). The evolution of automation in aviation, from the "gyroscopic autopilot" developed by the Sperry Corporation in 1912 to the state-of-the-art autonomous aircraft, has laid the groundwork for the next step toward fully automated AAM vehicles.

Emerging technologies, such as eVTOL vehicles and aircraft that will provide transportation services within the AAM concept of operations, have benefited from a century of technological innovations that have increased aircraft control while reducing the pilot's workload. However, in terms of aircraft control and automation, the AAM vision requires aircraft systems capable of moving from reduced human intervention to fully autonomous aircraft. In other words, the AAM vision requires aircraft that transition from pilot-centered or pilot-in-the-loop to fully autonomous aircraft capable of operating in dense airspace with minimal human supervision. Full autonomy on the eVTOL aircraft itself is one of the core operational enablers within the AAM vision. Automation is a scalability enabler. Autonomous eVTOL operations have the potential to support high-frequency, on-demand services that extend well beyond the limitations of pilot availability, human workload, and traditional ATM practices. Therefore, autonomy is one of the four operational enablers analyzed in this chapter.

Literature on automation capabilities in aviation encompasses a range of capabilities, perspectives, and human-centered approaches. The National Aeronautics and Space Administration (NASA, 1991) describes different levels of automation, each with specific capabilities or characteristics. Sheridan and Verplanck (1978) understand the concept of automation as the levels of control between human operators and machines. Another approach emphasizes *adaptive automation*, in which the levels of control vary depending on the context, workload, and operator's conditions (Parasuraman, 2001). In complex scenarios, including emergencies or unplanned events, *adaptive automation* enhances the level of control and support provided to the operator, enabling them to overcome unexpected events. Conversely, when the workload is low, automation capabilities are reduced, giving the operator an increased level of control during routine operations. A different approach to automation is provided by the Human Systems Integration (HIS) perspective. HIS automation design prioritizes human performance, needs, and operational limitations to balance control and supervision (Rippy, 2021). The HIS provides a broader framework where the "human" refers to different dimensions or roles during the operation, ranging from operators, trainers, maintainers, suppliers, and customers, among other actors. From the HIS perspective, flight deck automation is designed to support pilots during the aircraft operation, facilitate training requirements (Flight Simulator Training Devices [FSTD]), and support automated capabilities related to maintenance tasks. A broader approach during automation design incorporates the needs of trainers, maintainers, and other stakeholders, not just the operator's requirements in the cockpit. The HIS framework applied to automation aims to enhance the human-machine coordination and communication to the involved actors.

A Pilot Automation Interaction (PAI) framework developed by Feary et al. (2023) uses four levels to evaluate Urban Air Mobility (UAM) vehicles: (1) Stability Augmented Control, (2) Autopilot Command, (3) Task Management, (4) Mission Management. In the PAI framework, the authors describe initial levels of automation as vehicles limited to axis and vector control. Vehicles in the intermediate levels of automation are capable of controlling the aircraft in a set of maneuvers, flight path management, and managing tasks related to complex scenarios or responding to emergencies. On the PAI framework, fully autonomous aircraft are capable of overcoming operational contingencies (e.g., delays, congested airspace, and unexpected events) without human intervention. Rather than focusing on binary outcomes, automation in the AAM concept of operations is driven by the vehicle's performance and the type of operation (e.g., air taxi, Emergency Medical Services (EMS), logistics). In this chapter, the levels of automation for eVTOL operations range from *Level 0—Fully Piloted* to *Level 5—Full Automation*.

The automation framework of the AAM concept of operations deviates from the "all or nothing" condition, focusing on developing capabilities and managing different levels of control. Early AAM automated vehicles may use traditional automation capabilities (Levels 0, 1, and 2) where the eVTOL capabilities allow aircraft control, but the pilots remain central to the operation. Intermediate (Level 3) automation capabilities play a leading role during the execution of aircraft control, obstacle clearance, and trajectory management, while reserving the management of unexpected events/contingencies for the pilot's intervention. At high autonomy capabilities (Level 4), eVTOL vehicles can perform aircraft navigation, communication, and complete most tasks related to both normal and abnormal operations. The pilot's intervention on Level 4 of automation is limited to flying conditions within the scope of the system's design. Full autonomy (Level 5): all functions are managed by the automation capabilities. Conditions such as management of emergencies, delays, route optimization, and detection and avoidance are handled autonomously by the eVTOL vehicle. At this stage, any pilot's supervision or human-in-the-loop requirement is eliminated. Table 6.1 adapts this framework to the cruise phase of flight to describe the automation requirements for eVTOL vehicles.

The concept of intelligent flight controls introduced by Stengel (1993) describes aircraft control capabilities as a system capable of emulating human intelligence. An intelligent flight control system is capable of (1) reducing the need for human intervention, (2) improving aircraft control, (3) learning from previous flights, (4) increasing reliability, and (5) reducing the cost associated with operating the aircraft (Stengel, 1993). The contemporary developments and capabilities of AI applied to autonomous flight control systems represent one critical requirement for autonomous UAM operations (Berkol & Demirtaş, 2024). Unlike earlier generations of FMS and autopilots, *AI-Driven Flight Control Systems* use machine learning to enhance maneuvering capabilities and aircraft control. Current developments on Stability Augmentation Systems (SAS) have the potential to continue increasing automation adaptability. The Aircrew Labor In-Cockpit Automation System (ALIAS) was designed as a drop-in

TABLE 6.1
eVTOL Levels of autonomy and functional allocation—cruise phase of flight

Autonomy Level	Degree of Automation Support	Characteristics during Cruise Phase
Level 0 – Fully Piloted	100% Pilot (Navigation, FMS, Detect & Avoid, Comms)	Pilot manages all navigation, airspace separation, communications with ATC, and system monitoring. Limited to autopilot hold modes.
Level 1 – Assisted Flight	Shared Function (Navigation, Detect & Avoid, FMS)	AI assists with basic tasks: stability augmentation and route tracking. Pilot retains full authority. Comparable to SAE Level 1 driver assistance.
Level 2 – Partial Automation	Shared Function leaning toward Automation	AI handles lateral/vertical guidance and monitors traffic/weather. Pilot supervises and intervenes. Comparable to SAE Level 2.
Level 3 – Conditional Autonomy	Automation leads, pilot supervises (Human-on-the-loop)	AI executes cruise navigation, separation, and trajectory. Pilot intervenes if automation requests. Similar to SAE Level 3 driving.
Level 4 – High Automation	Mostly Automated (Navigation, Detect & Avoid fully automated)	Aircraft can complete cruise without human input under normal and most abnormal conditions. Pilot acts only in exceptional contingencies.
Level 5 – Full Automation	100% Automation (Navigation, Detect & Avoid, FMS, Comms)	No pilot involvement. AI manages all cruise functions, contingency rerouting, and ATC-equivalent interactions.

kit to enable high levels of automation in existing aircraft. ALIAS aims to reduce pilot workload, augment mission performance, and improve safety by supporting mission execution from takeoff to landing, as well as response to emergencies (DARPA, 2024). The ALIAS and Artificial Intelligence Reinforcement (AIR) programs demonstrated that neural-network-based flight control systems autonomously executed complex maneuvers and managed contingencies, such as responses to engine failures, validated through more than 200 hours of flight testing in helicopters and fixed-wing aircraft. State-of-the-art flight control systems use data from multiple sensors and navigation systems to anticipate the effect of adverse weather conditions on stability and aircraft control (Imran, 2025). These systems fuse inputs from GNSS, inertial sensors, radar, lidar, and electro-optical sources to generate state estimations and flight paths under degraded conditions (EASA, 2024).

The evolution of flight control technologies illustrates a century-long trajectory from mechanical stabilization to intelligent autonomy. Early autopilots and fly-by-wire systems reduced workload and improved safety but still required constant pilot oversight. Contemporary eVTOL and AAM concepts extend this legacy by integrating AI into flight control computers, enabling adaptive automation that can manage both routine operations and unexpected contingencies. NASA's levels of automation, HIS principles, and PAI frameworks recognize that autonomy is not binary but a spectrum in which human and machine roles are changing and adapting rapidly. Research trends in AI-driven flight controls emphasize (1) explainable AI (XAI) for certification, (2) resilient flight control laws capable of learning from in-service data, and (3) integration with digital twins (DTs) for continuous safety validation (EASA, 2024). In this trajectory, flight control computers evolve from deterministic systems into predictive, context-aware agents that not only stabilize aircraft but also aviate, optimize energy use, and manage contingencies without human intervention. Case studies, including DARPA's ALIAS program and Airbus's Project Vertex (Airbus, 2023), demonstrate that high levels of automation are feasible in both conventional and next-generation platforms. At the same time, AI-driven detect-and-avoid, stability augmentation, and trajectory management systems highlight the path toward scalable urban operations. The benefits of these capabilities are represented in a reduced pilot workload, enhanced safety margins, and operational scalability. Make autonomy a cornerstone and foundational step for the AAM integration. Yet, challenges remain in the certification of machine-learning-based systems, meeting explainability requirements, and cultivating trust among regulators, operators, and the public. Ultimately, AI-driven flight control is not only an enabler but also a prerequisite for realizing safe, efficient, and sustainable AAM.

6.3 PREDICTIVE MAINTENANCE AND AI DIAGNOSTICS

Gustavo A. Sanchez

Predictive maintenance (PdM) and AI diagnostics applied to the AAM concept of operations represent an innovation that departs from traditional maintenance practices. Applying AI capabilities to maintenance creates scenarios where potential failures are identified before their materialization. Predictive capabilities reinforced by the use of AI may reduce unscheduled maintenance and optimize the logistics required for the high volume of operations envisioned on the AAM. From an economic standpoint, reducing aircraft on ground (AOG) times, combined with efficiencies in inventory management and resource allocation (e.g., supplies, tools, and equipment), becomes an enabler for the success of the new way of transporting goods and services using eVTOL vehicles. Within the PdM scope, multiple technologies are being applied to the maintenance ecosystem. This section evaluates the potential role of DT, Integrated Vehicle Health Management (IVHM), AI-driven anomaly detection, and prognostics during the eVTOL operations.

A DT is defined as a virtual representation of a physical asset, equipped with data that mimics the structure and behavior of the physical asset throughout its life cycle. Historically, DT technologies have been useful for exploring potential weaknesses or failures of emerging technologies. In the aerospace sector, this concept is exemplified in complex operations, such as NASA's Apollo Program during the 1970s (Xiong & Wang, 2022). Engineers have used DT to improve the maintenance inspections and repair planning. A systematic review of DT, Bisanti et al. (2023) surveyed the state of DT applications in aviation and proposed an IoT-enabled modular architecture where sensor data and models (using data from the physical asset) interact continuously. Their analysis emphasized connectivity as a critical enabler and highlighted the lack of standardized frameworks for real-time synchronization. For eVTOL fleets, which will rely on constant data streaming from distributed sensors. A DT and the modular design offer a template for scalable PdM adoption within the AAM operations.

More advanced application of the DT concept to PdM focuses not only on anticipating failures but also on evaluating mission readiness. Kapteyn et al. (2021) presented a structural twin for an unmanned aerial vehicle that updated probabilistic models as degradation occurred in flight to optimize mission plans. Increased levels of reliability derived from the application of the DT technology to PdM may be beneficial for the eVTOL application in emergency response operations. Critical air transport missions, including EMS, response to natural disasters, and law enforcement operations, find in DT an enabler to the deployment of eVTOL vehicles that are continuously monitored to anticipate failures while transporting patients or providing lifesaving services. The DT technology applied to eVTOL operations may contribute to increasing readiness standards and improving public perception. Factors that have been identified as prerequisites for the adoption of eVTOL vehicles.

The IVHM provides another technology that may improve the benefits of PdM practices. Traditionally, maintenance practices assumed interval-based checks as the best alternative to prevent equipment failures. Departing from the traditional maintenance practices, IVHMS uses data from sensors distributed on the aircraft to improve the Condition-Based Maintenance (CBM) practices (IATA Maintenance Group, 2022). According to the study from IATA (2022), 89% of aircraft failures do not correlate with age or specific intervals, which reinforces the relevance of applying IVHM to eVTOL operations. IVHM uses data from different aircraft subsystems to understand the effect of failures on other subsystems (Ezhilarasu et al., 2019). The application of IVHM to commercial aviation enhances safety while reducing maintenance costs by (1) prioritizing maintenance, (2) increasing reliability, and (3) increasing the product life cycle, among other factors (Chang & Wang, 2017).

For eVTOL operations, IVHM represents a key enabler of mission readiness. Ensuring that eVTOLs can be dispatched with greater confidence in their health status while minimizing the risks of unexpected failures in AAM operations. The implementation of IVHM in commercial aviation operations has been combined with the benefits provided by AI applications in deep learning (DL) and machine learning (ML) (Kwakye et al., 2024). The combination of AI with IVHM for diagnostic purposes represents a deviation from traditional maintenance practices. When integrated with AI, these technologies not only identify issues but also assist with fleet management, mission planning, and regulatory compliance. These factors are crucial for the expansion of eVTOL services in densely populated urban areas. This IoT-AI synergy becomes more relevant for eVTOL and AAM vehicles due to the increased amount of operational data. However, its success may depend on fixing problems with data quality, regulatory integration, and standardization across the AAM industry.

Building on DT and IVHM, the next layer of innovation in the eVTOL's maintenance domain comes from *AI fault prognostics.* AI-driven anomaly detection and fault prognostics extend PdM by moving from simple fault identification to forecasting when and how failures might occur. A DL method known as Long Short-Term Memory Autoencoders (LSTM-AE) has demonstrated strong potential by learning patterns of "normal" system behavior and alerting to deviations as early signs of failure appear in a system. Ning et al. (2021) explored the application of LSTM-AE for fault detection and classification for maintenance purposes. The authors found that LSTM-AE models could successfully detect subtle system faults, including pressure sensor leaks, that traditional monitoring methods overlooked.

Other research emphasizes the importance of detecting rare but catastrophic events, such as thermal runaway in batteries or control system faults. Imbalances in data availability affect predictive modeling. Dangut et al. (2021) addressed this challenge by developing a hybrid ML model that can handle imbalanced data, achieving better accuracy in predicting rare failures than conventional approaches. Complementing these detection methods, prognostic tools aim to estimate the progression of faults and remaining useful life (RUL) of components. Berghout et al. (2022) introduced ProgNet, a DL model trained on engine degradation data to forecast damage propagation. The authors explored the transferability across engine types. Applied to eVTOL systems, similar prognostic models could monitor rotor or battery health, enabling just-in-time maintenance interventions. Together, these approaches highlight how AI not only improves anomaly detection precision but also enables proactive planning, ensuring that faults are mitigated before they evolve into critical failures. PdM capabilities contribute to accelerating the capabilities envisioned in the AAM concept of operations.

AI diagnostics and PdM are two examples of how technology, economics, and regulations may interact to increase safety and reliability. For AAM and eVTOL operations, their adoption could be crucial due to the current limited infrastructure to perform maintenance tasks. Figure 6.1 illustrates the relation between these emerging technologies and the benefit provided to PdM.

Technology	Potential Benefits
Digital Twin (DT) • Virtual representation of eVTOLs • Real-time health monitoring • Mission readiness for EMS & logistics	**Safety** **Reduced AOG** **Cost Efficiency** **operational Reliability** **Public Trust**
Integrated Vehicle Health Management (IVHM) • Data-driven Condition-Based Maintenance (CBM) • Subsystem integration & lifecycle extension • Enhanced dispatch reliability	
AI Fault Prognostics • Anomaly detection (rare failure, subtle deviations) • Remaining Useful Life (RUL) forecasting • Just-in-time maintenance interventions	

FIGURE 6.1 Enablers of predictive maintenance (PdM) in eVTOL operations.

Note: * AOG—Aircraft on Ground.

6.4 ENERGY SYSTEMS: BATTERY TECHNOLOGIES AND CHARGING INFRASTRUCTURE

Gustavo A. Sanchez

Power eVTOL operations under the AAM remain one of the biggest challenges in terms of physical infrastructure and battery performance. This section analyzes emerging battery technologies and requirements for charging infrastructure. Specifically, this section focuses on (1) battery technologies, (2) life cycle management, (3) charging versus battery swapping, (4) vertiport integration, and (5) sustainability and supply chain. Unlike conventional helicopters that use fossil fuels to power their operations, eVTOL vehicles rely on electricity to provide power during all phases of flight.

For eVTOL operations, battery technologies capture the attention of regulators, manufacturers, and potential operators. Lithium-ion (Li-ion) batteries are the preferred choice among OEMs in the near term due to their current availability, established supply chains that facilitate an increasing number of eVTOL operations, and proven performance. According to NASA's *Transformative Flight Work—Working Group 4,* Li-ion batteries are increasing battery density by approximately 3–5% per year (Doo et al., 2021). The author highlighted that improvements associated with the frequency of battery replacements may reduce the costs associated with the eVTOL operations by 50%. The battery-specific power performance, or in other words, the amount of power a battery can deliver per unit of mass, is another key characteristic for eVTOL vehicles.

For eVTOL operations, high specific power is essential to sustain vertical takeoff, hover, and rapid climb phases. Flight envelopes that differ from the cruise phase of flight require a higher amount of power from the eVTOL batteries. A study examining the maximum range of eVTOL vehicles under simulated operational scenarios revealed that hovering for 10 minutes can result in a 75% reduction in range (Hagag et al., 2021). The authors explored different configurations of eVTOL vehicles (e.g., Lift & Cruise, Tilt Rotor, and Vectored Thrust), finding that headwind conditions of 20 m/s may decrease the eVTOL range by up to 71%, depending on the specific configuration. Li et al. (2025) employed a multidisciplinary design optimization (MDO) framework to investigate energy consumption under various operational conditions. The authors found that optimizing cruise altitude and hover time, among other factors, may result in reductions of 11.4% in total energy requirements. In the battery power domain, several areas of study regarding battery capabilities require applied research that gains insights from real-world operations as OEMs advance in the certification process. Specifically, power demands during different segments of mission profiles of eVTOL vehicles require further research to understand the limitations of this technology (Dixit et al., 2024).

The batteries' life cycle and charging time are another key characteristic for the eVTOL operations. Fast and reliable charging facilitates a high volume of eVTOL operations, reducing turnaround times. In AAM, vertiports will function as mini "electric airports," and their viability depends on power quality, protection, and process. The FAA-commissioned analysis with the National Renewable Energy Laboratory (NREL) showed that eVTOL fast-charging capabilities may increase a facility's peak demand by factors of 5–10 (Solanki et al., 2023). The FAA's *Engineering Brief 105A Vertiport Design, Supplemental Guidance to Advisory Circular 150/5390-2D, Heliport Design*, emphasizes the importance of life cycle on eVTOL operations by introducing geometry, safety, and electrical provisions specific to eVTOL, including space planning for charging equipment and the possibility of battery-service operations (FAA, 2024). The National Academies' airport-centric AAM synthesis introduces a practical planning dimension; many early operations will stage from existing airports, providing authorities with a near-term pathway to deploy charging where utilities and safety systems already exist—while still requiring upgrades, operating procedures, and training (NASEM, 2023).

Turnaround time is an economic and scheduling lever in AAM, and energy replenishment is the pacing item. Fast charging minimizes handling complexity but confronts thermal management and grid-peak challenges; battery swapping minimizes aircraft dwell time but shifts complexity to logistics and ground safety. An analysis from the Government Accountability Office (GAO, 2022) explored the use of battery swapping procedures as an alternative to reduce turnaround times during eVTOL operations. EASA's *Prototype Technical Design Specifications for Vertiports* (2022) and FAA's (2024) guidance for vertiport design recognize battery swapping and established provisions for increasing safety during battery handling and storage. These specifications provide a starting point for the regulatory framework that guides the actions of the local authorities, infrastructure developers, and safety personnel and standardize the characteristics of the physical environment at vertiports and airports.

As the AAM industry and eVTOL operations scale, harmonized standards for both charging and swapping facilities may be essential to balance operational efficiency and the safety of air operations.

In the safety domain, using batteries to power the eVTOL vehicles is one of the significant certification challenges for OEMs. The European Aviation Safety Agency (EASA) has emphasized the thermal-runaway risks associated with Li-ion batteries in aviation contexts and the need for robust battery management, containment, and handling procedures (EASA, 2022). The infrastructure requirements for eVTOL operations involve not only areas related to power availability. The physical environment needs include safety considerations. Electrical fires and battery thermal hazards are two areas that have been widely studied. In Li-ion batteries, a thermal runaway may result in fire or explosion of the battery (Wang et al., 2019). The lack of adequate fire suppression systems to extinguish an electrical fire at airports and vertiports highlights a critical gap in the current aeronautical infrastructure. Therefore, certification frameworks for eVTOL vehicles require protocols and specifications focused on mitigating the effect of electrical fires and battery thermal hazards: (1) fire suppression systems, (2) redundant battery monitoring and management systems, (3) containment structures, (4) battery handling procedures, and (5) training for the emergency response personnel at airports and vertiports. Table 6.2 summarizes the key battery performance characteristics that define suitability for eVTOL operations, reflecting both the technical thresholds and safety considerations required to support scalable eVTOL operations.

Sustainability in eVTOL energy systems is less about environmental goals and more about whether fleets can be supplied, maintained, and disposed of responsibly

TABLE 6.2
Key battery performance characteristics for eVTOL operations

Characteristic	Target for eVTOL Operations	Rationale
Specific Energy (Wh/kg)	≥400 Wh/kg*	Higher specific energy is critical—Extend range and maximize payload capacity.
Specific Power (W/kg)	>2,000 W/kg*	High power output—Vertical takeoff, hover, and rapid climb phases.
Cycle Life	>2,000 cycles at high C-rates	Frequent daily missions demand—Battery replacement costs.
Charge Time	<15–20 minutes (fast charge)	Minimizes turnaround time at vertiports, supporting high-frequency operations.
Thermal Stability	Robust containment	Mitigates the risk of thermal runaway.
Operating Temperature Range	–20°C to +60°C	Ensures performance in diverse environments and climates.
Safety Features	Advanced BMS, fault isolation, redundancy	Compliance with aviation certification standards.
Life cycle Sustainability	Recyclable materials	Reduces life cycle costs and environmental impact.

at scale. The use of batteries to power the aircraft introduces three interconnected sustainability challenges: *material availability, life cycle turnover, and infrastructure readiness*. Material *availability* refers to the levels of scarcity that each of the elements (e.g., lithium, nickel, and cobalt) presents in national and international supply chains, as well as the price volatility and social impact of mining practices. Additionally, the *life cycle turnover* refers to the potential accelerated wear of eVTOL batteries under specific mission profiles. A study from the Oak Ridge National Laboratory (ORNL, 2024) estimates that eVTOL duty cycles accelerate battery degradation compared to cars, necessitating more frequent battery replacements. The accelerated life cycle turnover evidences the importance of establishing recycling protocols and second-life applications for eVTOL batteries. Similarly, a robust charging infrastructure and grid capacity are fundamental capabilities to support the high volume of eVTOL operations.

The performance of eVTOL batteries must be evaluated not only by their energy capacity but also by their ability to meet the requirements of different mission profiles and contingencies proper to aviation operations. Industry and regulatory studies indicate that specific energy and specific power metrics have a direct impact on payload, range, and mission reliability during eVTOL operations. Emerging hazards related to using Li-ion batteries to power eVTOL vehicles, such as electrical fires and battery thermal runaways, pose systemic safety challenges that necessitate technological, regulatory, and training approaches to mitigate the impact of these safety events during air transport operations. Evaluating the sustainability of energy systems supporting a high volume of eVTOL vehicles goes beyond analyzing the environmental impact. The availability of the materials required to manufacture the batteries and the associated accelerated life cycle turnover are areas that need attention to reduce inefficiencies while exploiting the benefits of this emerging technology.

6.5 AIRSPACE MANAGEMENT AND AI-POWERED UTM SYSTEMS

Gustavo A. Sanchez

The AAM concept of operations expects a high volume of vehicles operating at low altitude. Increasing the number of operations in an already congested airspace requires new technologies to provide safe and efficient ATC services. The UAS Traffic Management (UASTM) is conceived as a cooperative structure where stakeholders (e.g., drone operators, service providers, and regulators) share data, responsibilities, and services to ensure an efficient use of the airspace. *As the ecosystem matures, the FAA will provide real-time constraints to the UAS operators, who are responsible for managing their operations safely within these constraints without receiving positive air traffic control services from the FAA* (FAA, 2023a).

The FAA's UTM Concept of Operations v2.0 in the United States formalizes a federated, community-based model. Under FAA oversight, third-party providers offer services like (1) flight planning, (2) intent sharing, (3) strategic deconfliction, and conformance monitoring. The FAA sets the performance requirements and operational policy (FAA, 2020). The *AAM Concept of Operations Version 2.0* (FAA, 2023a) expands the 2020 vision to include Beyond Visual Line of Sight (BVLOS) scenarios and operations in or near controlled airspace. It also introduces new concepts, such as volume reservations and performance authorizations, and explains how to manage emergencies between UTM and ATM. Making UTM a part of the National Airspace System (NAS) instead of a separate system (FAA, 2020).

In Europe, the counterpart to UTM is called U-space. It is not a single system operated by Eurocontrol or EASA but rather a set of digital services and rules that enable drones, and eventually eVTOL aircraft, to fly in very low-level (VLL) airspace (typically below 500 ft). The concept was defined in the Single European Sky ATM Research (SESAR) Joint Undertaking's U-space ConOps (CORUS), which describes U-space as a service-based ecosystem: private companies (called U-space service providers) deliver functions like flight planning, tracking, and conflict resolution, while the State regulates, accredits, and oversees them (SESAR, 2019).

U-space is built as a stepwise system of services.

- U1 (foundation services): basic functions such as drone registration and geo-awareness (knowing where no-fly zones are).
- U2 (initial services): flight planning, flight approval, and live tracking of drones.
- U3 (advanced services): strategic deconfliction, dynamic re-routing, and interface with air traffic control.
- U4 (full integration): high levels of automation, complex operations, and seamless interaction with the ATM.

The layered U-space design means that as traffic density and risk increase, more advanced U-space services are activated. For example, a rural inspection drone may only need U1 or U2, while dense urban eVTOL traffic would require U3 or U4 services (SESAR, 2019). For eVTOL operations, the UTM approaches in Europe and the U.S. are shifting routine separation and flight coordination tasks to a digital, federated UTM system, where regulators can ensure that eVTOLs operate safely and efficiently without overloading human controllers (FAA, 2020, 2023b).

The large number of operations envisioned on AAM demands new technologies for ATM. The use of AI applications to address the limitations of ATM infrastructure presents promising scenarios. European demonstrations concluded that advanced conflict detection is essential when multiple flights operate concurrently. In DOMUS, a federated U-space network, coordinated real-time deconfliction and dynamic rerouting with reaction times on the order of seconds, implying algorithmic optimization under the hood (SESAR, 2019). The GOF trials established an interoperability architecture that linked manned ATC systems and U-space via an open Flight Information Management System (FIMS), sharing real-time traffic data with all actors, which enabled mixed operations while exposing technical seams to resolve (e.g., altitude references) (SESAR, 2019).

NASA's Campaign (NC-1) "X4" built a UTM-inspired UAM prototype with multiple Providers of Services for UAM (PSUs) connected through a Discovery and Synchronization Service (DSS) and added Demand-Capacity Balancing (DCB) for urban corridors (Cheng et al., 2022). Seven industry teams connected their PSU prototypes and iterated through increasingly complex scenarios. Lessons directly inform AI's role: UAM services require (1) richer intent sharing, (2) robust conformance monitoring, and (3) dynamic replanning to handle higher speeds and densities; the DSS struggled as complexity scaled, signalling the need for more scalable data-exchange mechanisms and smarter coordination (Cheng et al., 2022). The experiment also piloted community-based priority rules among operators, using procedural constructs that algorithmic agents can implement consistently once codified (Cheng et al., 2022).

Despite improvements in both the U.S. and Europe in UTM and AI-enabled services, broader implementation requires efforts from stakeholders. Initially, standards and interoperability remain incomplete. Preliminary assessments (GOF and DOMUS) revealed that some providers used different benchmarks for altitude, data formats, and latency (SESAR, 2019). This renders AI systems less dependable due to their requirement for high-quality data and consistent inputs. Secondly, there remain deficiencies in governance. The GAO and DOT studies emphasize that the FAA has yet to establish definitive milestones, performance criteria, and delineated roles for several agencies, hence complicating industry investments in AI technologies (DOT, 2023; GAO, 2021). Eurocontrol determined that several European countries were unprepared, with several remaining at U1–U2 levels and dependent on private providers whose data-sharing protocols require stringent oversight (Eurocontrol, 2022).

Ensuring the safety of AI operations presents a distinct challenge. The National Academies and EASA assert that AI must be explicable, resilient to cyberattacks, and capable of managing erroneous data before certification for aviation applications. Ultimately, public trust remains tenuous. To develop it, challenges such as noise, privacy, cybersecurity, and equitable access must be addressed through transparent governance, communications, and the appropriate application of AI (NASEM, 2022; WEF, 2025). The AI-driven UTM cannot achieve the requisite scale or legitimacy for eVTOL integration unless these issues are resolved.

6.6 CONCLUSION

Dimitrios Ziakkas, Debra Henneberry, and Amin Vafadar

A century of automation has progressively transferred routine stabilization, guidance, and optimization from the pilot's hands to computational agents; AAM extends this logic into dense, time-critical, low-altitude operations where consistency, responsiveness, and coordination are paramount. The evidence base shows that AI-enabled control laws, fed by multi-sensor fusion and bounded by rigorous protections, can navigate both nominal and off-nominal conditions, from gust-induced upsets to subsystem degradations. Yet the human remains central. Competence now resides as much in supervising adaptive systems, managing intent with external services, and arbitrating edge cases as in manual control. In practice, safety emerges from the alignment of three layers: pilot-automation teaming that is transparent and predictable; certifiable AI that is explainable, robust to cyber-physical perturbations, and continually validated against operational data; and airspace-level orchestration (UTM/U-space) that supplies the external constraints within which autonomous agents plan and replan.

Realizing this alignment requires disciplined sequencing. Early service models will sensibly anchor at "conditional" and "high" automation with human-on-the-loop authority, leveraging simulator-rich training to surface workload, mode-awareness, and surprise-management issues before high-energy flight. As evidence accumulates—from structured flight-test campaigns, in-service monitoring, and digital-twin analyses—functions can migrate toward mission-level autonomy with explicitly defined performance envelopes and fallbacks. Certification practice will have to absorb not only deterministic requirements but also data-centric assurance cases that track how learning systems behave over time; explainable AI and continuous airworthiness monitoring will be the connective tissue between laboratory promise and public trust (EASA, 2024).

The payoff justifies the discipline. Autonomy, properly integrated, is the mechanism by which eVTOL services scale beyond the constraints of pilot availability and human attention, delivering repeatable trajectory management, energy-aware decision-making, and resilient contingency handling at urban tempo. Its success, however, is inseparable from the people who design, certify, operate, and maintain it. When automation is shaped around human strengths and limitations—and embedded in transparent procedures, humane training pathways, and accountable governance—it does more than fly the aircraft: it legitimizes AAM as a safe, efficient, and sustainable public service.

REFERENCES

Abzug, M. J., & Larrabee, E. E. (2002). *Airplane stability and control: A history of the technologies that made aviation possible* (2nd ed.). Cambridge University Press.

Airbus. (2023, November 22). *At your fingertips: Airbus flies a fully automated helicopter with a tablet.* https://www.airbus.com/en/newsroom/press-releases/2023-11-at-your-fingertips-airbus-flies-a-fully-automated-helicopter-with-a-tablet

Berghout, T., Bentrcia, T., Ferrag, M. A., & Benbouzid, M. (2022). A heterogeneous federated transfer learning approach with extreme aggregation and speed. *Mathematics*, *10*(19), 3528. https://doi.org/10.3390/math10193528

Berkol, A., & Demirtaş, İ. G. (2024, December 22). *Autonomous flight systems and generative AI.* https://dergipark.org.tr/en/pub/ijmsit/issue/88457/1583035

Bisanti, G. M., Mainetti, L., Montanaro, T., Patrono, L., & Sergi, I. (2023). Digital twins for aircraft maintenance and operation: A systematic literature review and an IoT-enabled modular architecture. *Internet of Things*, *24*, 100991. https://doi.org/10.1016/j.iot.2023.100991

Boeing. (2023, February). *Airplane characteristics for airport planning, revision E (Document No. D6–86073).* Boeing. https://www.boeing.com/content/dam/boeing/boeingdotcom/commercial/airports/acaps/777X_RevE.pdf

Chang, Shuo, & Wang, Yi. (2017). Integrated vehicle health management technology and its applications in commerical aviation. *2017 International Conference on Sensing, Diagnostics, Prognostics, and Control (SDPC)*, 740–745. https://doi.org/10.1109/SDPC.2017.145

Cheng, A. W., Witzberger, K. E., Isaacson, D. R., Verma, S. A., Arneson, H. M., Lee, H., Seah, C. H., & Levitt, I. (2022). *National Campaign (NC)-1 strategic conflict management simulation (X4) final report* (NASA/TM-2022-0018159). NASA Ames Research Center; NASA Langley Research Center. https://ntrs.nasa.gov/citations/20220018159

Dangut, M. D., Skaf, Z., & Jennions, I. K. (2021). An integrated machine learning model for aircraft components rare failure prognostics with log-based dataset. *ISA Transactions*, *113*, 127–139. https://doi.org/10.1016/j.isatra.2020.05.001

David, A. (2024). *Controlling aircraft: From humans to autonomous systems: Rise of the machines*. https://doi.org/10.4271/EPR2024020

Defense Advanced Research Projects Agency (DARPA). (2024). *ALIAS: Aircrew labor in-cockpit automation system*. DARPA. https://www.darpa.mil/research/programs/aircrew-labor-in-cockpit-automation-system

Dixit, M., Bisht, A., Essehli, R., Amin, R., Kweon, C. M., & Belharouak, I. (2024). Lithium-Ion battery power performance assessment for the climb step of an electric vertical Takeoff and landing (EVTOL) application. *ACS Energy Letters*, *9*(3), 934–940. https://doi.org/10.1021/acsenergylett.3c02385

Doo, J. T., Tsairides, M. A., Pavel, M. D., et al. (2021). *NASA electric vertical takeoff and landing (eVTOL) aircraft technology for public services—A white paper (Working Group 4)* (NASA Document No. 20205000636). NASA Transformative Vertical Flight. https://ntrs.nasa.gov/api/citations/20205000636/downloads/2021-08-20-eVTOL-White-Paper-Final_V48.pdf

Enenakpogbe, E., Whidborne, J. F., & Lu, L. (2024). Full envelope control of over-actuated fixed-wing vectored thrust eVTOL. *Aerospace*, *11*(12), 979. https://doi.org/10.3390/aerospace11120979

EUROCONTROL. (2022). *U-space services implementation monitoring report* (Edition 2022). EUROCONTROL. https://www.eurocontrol.int/sites/default/files/2022-11/eurocontrol-u-space-services-implementation-monitoring-report-2022.pdf

European Union Aviation Safety Agency. (2022). *Prototype technical design specifications for vertiports*. EASA. https://www.easa.europa.eu/en/downloads/136946/en

European Union Aviation Safety Agency (EASA). (2024, April 19). *EASA artificial intelligence concept paper: Issue 2—Guidance for Level 1 & 2 machine-learning applications*. https://www.easa.europa.eu/en/document-library/general-publications/easa-artificial-intelligence-concept-paper-issue-2

Ezhilarasu, C. M., Skaf, Z., & Jennions, I. K. (2019). The application of reasoning to aerospace Integrated Vehicle Health Management (IVHM): Challenges and opportunities. *Progress in Aerospace Sciences*, *105*, 60–73. https://doi.org/10.1016/j.paerosci.2019.01.001

Feary, M. S., Kaneshige, J., Lombaerts, T., Shish, K., & Haworth, L. (2023, June 12). *Evaluation of novel EVTOL aircraft automation concepts*. NASA Technical Reports Server (NTRS). https://ntrs.nasa.gov/citations/20230009729

Federal Aviation Administration [FAA]. (2020). *Urban air mobility concept of operations v1.0*. U.S. Department of Transportation. https://nari.arc.nasa.gov/sites/default/files/attachments/UAM_ConOps_v1.0.pdf

Federal Aviation Administration [FAA]. (2023a). *Advanced Air Mobility (AAM) implementation plan*. https://www.faa.gov/sites/faa.gov/files/AAM-I28-Implementation-Plan.pdf

Federal Aviation Administration [FAA]. (2023b). *Unmanned Aircraft System (UAS) Traffic Management (UTM)*. U.S. Department of Transportation. https://www.faa.gov/uas/advanced_operations/traffic_management

Federal Aviation Administration [FAA]. (2023c). *Urban air mobility concept of operations 2.0*. U.S. Department of Transportation. https://www.faa.gov/sites/faa.gov/files/Urban-Air-Mobility-Concept-of-Operations-2.0.pdf

Federal Aviation Administration [FAA]. (2024). *Engineering brief No. 105A: Vertiport design (supplement to AC 150/5390-2D Heliport design)*. FAA Airports Division. www.faa.gov/airports/engineering/engineering_briefs/eb_105a_vertiports

Goldfeld, D. A., Mayer, L. A., Brown, J., Cochran, S. T., Hastings Roer, E., Litterer, S., Mason, R., Mignano, J. R., McBirney, S., Villegas, C. A., Project Air Force, Rand Corporation, &

United States. Department of the Air Force. (2024). *Amping airpower: Electric vertical takeoff and landing for the U.S. air force military utility, market dynamics, and warfighter adoption*. RAND. https://doi.org/10.7249/RRA1524-2

Hagag, N., Toepsch, F., Graf, S., Büddefeld, M., & Eduardo, H. (2021). Maximum total range of eVTOL under consideration of realistic operational scenarios. *Proceedings of the 11 th SESAR Innovation Days, Virutal*, 7–9.

Imran, L. J. (2025). Drone airflow dynamics and crop interaction. In *UAV aerodynamics and crop interaction. Smart agriculture* (Vol. 13). Singapore: Springer. https://doi.org/10.1007/978-981-96-8402-1_4

International Air Transport Association [IATA]. (2022). *From aircraft health monitoring to Aircraft Health Management (AHM): White paper* (1st ed.). IATA. https://www.iata.org/contentassets/fafa409c883d41198aeb87628c848851/ahm-wp-1sted-2022.pdf

Kapteyn, M. G., Pretorius, J. V. R., & Willcox, K. E. (2021). A probabilistic graphical model foundation for enabling predictive digital twins at scale. *Nature Computational Science*, *1*(5), 337–347. https://doi.org/10.1038/s43588-021-00069-0

Kwakye, A. D., Jennions, I. K., & Ezhilarasu, C. M. (2024). Platform health management for aircraft maintenance: A review. *Proceedings of the Institution of Mechanical Engineers. Part G, Journal of Aerospace Engineering*, *238*(3), 267–283. https://doi.org/10.1177/09544100231219736

Li, W., Cheng, R., Huang, H., Garg, A., & Gao, L. (2025). Energy consumption modeling and optimization of an EVTOL aircraft: Integrating weight, motor, and battery dynamics. *Energy*, 136229. https://doi.org/10.1016/j.energy.2025.136229

Liu, Y., Druyor, C. T., & Wang, L. (2023). High-fidelity analysis of Lift+Cruise VTOL urban air mobility concept aircraft. In *AIAA AVIATION 2023 forum*. American Institute of Aeronautics and Astronautics.

Malpica, C., et al. (2023, May). *Handling qualities of multirotor RPM-controlled electric vertical takeoff and landing (eVTOL) aircraft for urban air mobility (UAM)*. 79th Annual Forum & Technology Display (Forum 79), Vertical Flight Society, West Palm Beach, FL.

National Academies of Sciences, Engineering, and Medicine [NASEM]. (2022). *Preparing your airport for electric aircraft and hydrogen technologies*. Washington, DC: The National Academies Press. https://doi.org/10.17226/26512

National Academies of Sciences, Engineering, and Medicine [NASEM]. (2023). *Airport-centric Advanced Air Mobility market study*. Washington, DC: The National Academies Press. https://doi.org/10.17226/27326

National Academies of Sciences, Engineering, and Medicine [NASEM]. (2024). *Advanced Air Mobility and community outreach: A primer for successful stakeholder engagement*. Washington, DC: The National Academies Press. https://doi.org/10.17226/27627

National Aeronautics and Space Administration [NASA]. (1991). *Human-centered aircraft automation: A concept and guidelines* (Vol. 103885). National Aeronautics and Space Administration, Ames Research Center.

Ning, S., Sun, J., Liu, C., & Yi, Y. (2021). Applications of deep learning in big data analytics for aircraft complex system anomaly detection. *Proceedings of the Institution of Mechanical Engineers Part O Journal of Risk and Reliability*, *235*(5), 923–940. https://doi.org/10.1177/1748006x211001979

Oak Ridge National Laboratory [ORNL]. (2024, March 12). *More than flying cars: eVTOL battery analysis reveals unique operating demands*. https://www.ornl.gov/news/more-flying-cars-evtol-battery-analysis-reveals-unique-operating-demands

Parasuraman, R. (2001). *Adaptive automation and pilot performance* (NASA Contractor Report No. NASA/CR-2001-211048). National Aeronautics and Space Administration. https://ntrs.nasa.gov/citations/20010029669

Rippy, L. O. (2021). *NASA human systems integration handbook.* National Aeronautics and Space Administration Technical Reports Server (NTRS). https://ntrs.nasa.gov/citations/20210010952

Semanjski, I., Pratelli, A., Pieraccini, M., Semanjski, S., Petri, M., & Gautama, S. (Eds.). (2023). *Urban Air Mobility (UAM).* MDPI. https://doi.org/10.3390/books878-3-0365-9162-9

SESAR Joint Undertaking. (2019). *Initial view on principles for U-space architecture* (Edition 01.04). SESARJU. https://www.sesarju.eu/sites/default/files/documents/u-space/SESAR%20principles%20for%20U-space%20architecture.pdf

Sheridan, T. B., & Verplanck, W. L. (1978). *Human and computer control of undersea teleoperators.* Technical Report. Cambridge, MA: MIT Man-Machine Laboratory.

Solanki, B., Sanders, P., Miller, E., Paudyal, P., Rathod, B., Abraham, S. A., Young, M., Avelino, A. F. T., Padullaparti, H. V., Cary, S., Hallock, C., Moriarty, K., Ellwood, G., Wang, J., Flores-Espino, F., Rane, J., Markel, T., Sanghvi, A., & National Renewable Energy Laboratory. (2023). *Federal Aviation Administration vertiport electrical infrastructure study.* Technical Report No. NREL/TP-5R00–86245, Rev. December 2023. National Renewable Energy Laboratory. https://www.nrel.gov/docs/fy24osti/86245.pdf

Stengel, R. F. (1993). Toward intelligent flight control. *IEEE Transactions on Systems, Man, and Cybernetics, 23*(6), 1699. https://stengel.mycpanel.princeton.edu/TIFC.pdfSu, J., Huang, H., Zhang, H., Wang, Y., & Wang, F.-Y. (2024). eVTOL performance analysis: A review from control perspectives. *IEEE Transactions on Intelligent Vehicles, 9*(5), 4877–4889. https://doi.org/10.1109/TIV.2024.3387405

Ugwueze, O., Statheros, T., Bromfield, M. A., & Horri, N. (2023, January 19). Trends in eVTOL aircraft development: The concepts, enablers, and challenges. In *AIAA SCITECH 2022 forum.* American Institute of Aeronautics and Astronautics.

U.S. Department of Transportation [DOT]. (2023). *DOT OIG: Mission and organization.* Office of Inspector General. https://www.oig.dot.gov/

U.S. Government Accountability Office [GAO]. (2021). *High-risk series: Dedicated leadership needed to address major management challenges and ensure accountability.* (GAO-21-119SP). https://files.gao.gov/reports/GAO-21-119SP/index.html

U.S. Government Accountability Office [GAO]. (2022). *Transforming aviation: Considerations for financing and other challenges for advanced air mobility.* (GAO-23-105188). https://www.gao.gov/assets/gao-23-105188.pdf

Wang, Z., Yang, H., Li, Y., Wang, G., & Wang, J. (2019). Thermal runaway and fire behaviors of large-scale lithium ion batteries with different heating methods. *Journal of Hazardous Materials, 379*, 120730. https://doi.org/10.1016/j.jhazmat.2019.06.007

World Economic Forum. (2025). *Global aviation sustainability outlook 2025.* https://reports.weforum.org/docs/WEF_Global_Aviation_Sustainability_Outlook_2025.pdf

Xiong, M., & Wang, H. (2022). Digital twin applications in aviation industry: A review. *The International Journal of Advanced Manufacturing Technology, 121*(9–12), 5677–5692. https://doi.org/10.1007/s00170-022-09717-9

7 Stakeholder Collaboration and Global Standardization in AAM

Dimitrios Ziakkas, Amin Vafadar, and Debra Henneberry

7.1 INTRODUCTION

Advanced Air Mobility is a systems enterprise rather than a single product launch. Electric vertical take-off and landing aircraft carry with them an appealing promise—quieter, shorter, and point-to-point journeys across complex urban and regional geographies—but the promise materializes only when design assurance, operational approval, infrastructure, and public license move in step. The recent evolution of regulatory frameworks on both sides of the Atlantic shows that authorities increasingly treat AAM as a civil transport system that must be integrated with existing air traffic management and city mobility plans, rather than an isolated aviation novelty. In Europe, EASA's Innovative Air Mobility package and its integration with U-space make plain that piloted VTOL operations will begin within the established ATM/ATS architecture while steadily adding digital services and performance assurances that allow density to rise. In the United States, the FAA's Innovate28 program and the powered-lift rule establish the legal basis for training, licensing, and operations, while sequencing key-site deployments that can be audited, iterated, and scaled.

Within this framing, stakeholders assume distinct yet interdependent roles. OEMs shoulder certification and human-system integration obligations; operators must build safety management capacity, competency-based training pipelines, and community-facing practices; and regulators curate the enabling architecture and the evidentiary burden that underwrites public confidence. The connective tissue is standardized: transatlantic procedures for validation, common vertiport specifications, human factors and HMI guidance, and telecom-aviation agreements for data links. International organizations anchor these threads. ICAO defines the lexicon and the baseline methods needed for portability of approvals and personnel licenses; EUROCAE, RTCA, ASTM, and ISO translate those principles into interoperable artifacts for aircraft, infrastructure, and digital services. As demonstrations in Rome, Paris, Dubai, and Singapore confirm, credible progress arises where these elements are jointly orchestrated and where measurable milestones—first vertiport works, first local aircraft, and first controlled revenue trials—are accompanied by transparent engagement with residents and local authorities. The chapter that follows examines

DOI: 10.1201/9781003685555-7

this orchestration: the stakeholder compact, the harmonization lattice, the portability of training and certification, the policy instruments that anchor public trust, and the regional case studies that convert standards into practice.

7.2 STAKEHOLDERS IN THE AAM-eVTOL ECOSYSTEM: OEMS, OPERATORS, AND REGULATORS

Amin Vafadar

Advanced Air Mobility (AAM) is not a single-actor endeavor; its safe and scalable introduction depends on tightly coupled actions among original equipment manufacturers (OEMs), prospective operators, and regulators—supported by air navigation service providers (ANSPs), vertiport developers, municipalities, utilities, and standards bodies. The European Union Aviation Safety Agency (EASA) has explicitly begun codifying the regulatory substrate for "Innovative Air Mobility" (IAM), introducing a dedicated Annex (Part-IAM) and tailored guidance under Air OPS to frame operations, pre-flight preparation, energy management, and vertiport availability—signalling a civil, operations-first pathway in Europe (EASA, 2025). In the United States, the Federal Aviation Administration's Innovate28 (I28) Implementation Plan sets out a phased path to initial entry-into-service at key sites before scaling to a mature state, and—crucially—locates that trajectory within a multi-stakeholder compact spanning certification, airspace/ATC, infrastructure, environment, hazardous materials safety, and community engagement (FAA, 2023b).

- **OEMs:** AAM OEMs (e.g., Joby, Archer, and BETA) shoulder airworthiness, systems safety, and human-system integration responsibilities while demonstrating operational concepts that inform regulators' rulemaking and acceptable means of compliance. In EASA's jurisdiction, that means aligning design, crew licensing, and operations with evolving IAM provisions and extant aircrew and FSTD frameworks; in the United States, it means sequencing type certification, production certification, and operational approvals under the I28 roadmap (EASA, 2025; FAA, 2023a). Strategically, leading OEMs have also entered airline and infrastructure partnerships that accelerate ecosystem readiness—illustrated by Joby's multiagency program in Dubai with the Roads and Transport Authority (RTA), the Dubai Civil Aviation Authority (DCAA), and the UAE General Civil Aviation Authority (GCAA), including a six-year exclusive operating agreement and a four-vertiport starter network (Joby Aviation, 2024a, 2024b, 2025).
- **Operators:** Future AAM operators will likely comprise airline-affiliated ventures, helicopter incumbents diversifying into eVTOL fleets, and new entrants. Their accountabilities span flight operations, safety management, crew competence (including competency-based training), and community/environmental stewardship. The FAA's I28 plan explicitly states that operators and manufacturers must secure necessary certifications/approvals, engage affected communities, and minimize local impacts—placing public acceptance alongside regulatory compliance (FAA, 2023b, pp. 7–8). Airline-anchored models are already crystallizing, for example, Archer-United's announced airport-city

route in Chicago (ORD-Vertiport Chicago) and associated pre-delivery commitments; similarly, Joby-Delta's home-to-airport concept in New York and Los Angeles demonstrates an integrated multimodal offer and demand-aggregation logic (Delta Air Lines, 2022). In Europe, operator readiness must be synchronized with Part-IAM and Air OPS changes as EASA moves from demonstration to controlled service paradigms.
- **Regulators:** Regulators provide the enabling—and constraining—architecture. EASA's IAM regulatory package, combined with guidance on vertiports, energy planning, and operational contingencies, indicates a system view that integrates air and ground elements (EASA, 2025). The FAA's I28, meanwhile, articulates a "crawl-walk-run" approach to permit early operations with minimal disruption to existing services and infrastructure while developing advanced capabilities and procedures for scale (FAA, 2023a, §1.2–§3). Above the regional layer, ICAO's AAM Study Group has been convened to craft a holistic, globally harmonized framework and to route specific tasks to technical panels for Standards and Recommended Practices (SARPs)—a prerequisite for international interoperability and cross-border services (FAA, 2023a, pp. 6–7).

Although OEMs, operators, and regulators form the nucleus, several additional stakeholders are structurally indispensable to civil AAM:

- ANSPs and U-space/UTM actors. European U-space implementation (phased U1–U4) brings ANSPs, U-space service providers, cities, and regulators into a single operational fabric; Eurocontrol's pan-European networks and SESAR's CORUS-XUAM ConOps institutionalize this coordination and feed EASA's regulatory work (Eurocontrol, 2022, 2024; SESAR, 2023).
- Infrastructure developers and utilities. Vertiport developers (e.g., Skyports, UrbanV, and Groupe ADP/RATP) and electric-power providers underpin network viability. FAA vertiport engineering circulars and ASTM F3423/F3423M give initial, civil-use design baselines, while city/regional authorities manage land use, zoning, and power-distribution upgrades (ASTM, 2022; FAA, 2022; FAA, 2023a, pp. 6–7). Concrete tripartite agreements—such as RTA-Skyports-Joby in Dubai—illustrate how infrastructure, operator, and government can be contractually aligned for launch (Skyports, 2024).
- Standards bodies. RTCA and EUROCAE's renewed memorandum of cooperation accelerates joint, interoperable standards (e.g., avionics and CNS/ATM integration), reducing duplication and enabling transatlantic alignment that EASA and FAA can reference in guidance material.

From an ecosystem perspective, the literature is clear that "going it alone" is not viable: commercialization depends on consortia that marry airworthiness and operations with infrastructure, standards, and public-sector mandates (Roland Berger, 2018, 2020). In practice, this means synchronizing OEM certification milestones with operator training/operational approvals, vertiport readiness, U-space/ATC procedures, and public-engagement programs—under the supervisory logic of EASA/FAA and with ICAO harmonization for international scalability (Table 7.1).

TABLE 7.1
Stakeholder collaboration and global standardization in AAM

Stakeholder	Primary Roles in AAM	Near-term Deliverables (2024–2026)	Key Dependencies	Illustrative References
OEMs (e.g., Joby, Archer, BETA)	Airworthiness; systems safety; HMI; data packages for training/FSTDs; operational demos that inform rulemaking.	Type & production certification; safety cases aligned to EASA IAM/FAA SFAR & I28; airline/infrastructure partnerships; training syllabi; & FSTD data.	Timely certification guidance (EASA SC-VTOL/IAM; FAA powered-lift; & I28); vertiport readiness; U-space/ATM procedures; power/charging availability.	EASA (2025); FAA (2023a); Joby-RTA Dubai program (2024–2025).
Operators (airline-affiliated, helicopter incumbents, new entrants)	Flight ops & SMS; crew licensing/CBTA; environmental & community stewardship; integration with airports/metro mobility.	Ops specs/LoAs; initial route proving; community engagement plans; safety performance monitoring; intermodal offer (airport connectors).	OEM delivery & training; regulator approvals; vertiport slots/energy; ANSP/USSP interfaces; local land use permissions.	FAA I28 (2023b); Part-IAM EASA (2025); Archer-United ORD-city route (2023); Delta-Joby concept (2022).
Regulators (EASA, FAA; ICAO AAM-SG)	Policy, rules & AMC/GM; licensing frameworks; validation pathways; global harmonization.	EASA IAM package & U-space alignment; FAA powered-lift SFAR implementation; TIP Rev.7 + Amd 1 workflows; ICAO AAM lexicon/SARPs path.	Evidence from demos; SDO outputs; State & city coordination; international acceptance (BASA/TIP).	EASA (2025); FAA (2023a, 2024c); ICAO AAM-SG.
ANSPs & U-space/UTM actors	Procedural and digital integration (ATM↔U-space); strategic deconfliction; traffic information; conformance monitoring.	ATS-USSP protocols; initial corridors near TMAs; service-level monitoring.	Common Information Services; data quality & latency; city/U-space designation; telecom QoS.	EUROCONTROL; SESAR CORUS-XUAM (2023).
Infrastructure developers & utilities (Skyports, UrbanV, airport sponsors)	Vertiport siting/design/ops; energy systems; MRO & passenger processing; intermodal integration.	Early vertiports at airports/intermodal nodes; ED-308-ready charging; NFPA-aligned ESS rooms; safety & emergency plans.	ASTM/FAA/ISO guidance; grid upgrades; planning approvals; noise policy & community buy-in.	ASTM F3423; FAA EB-105/105A (2022/2024).
Standards bodies (EUROCAE, RTCA, ASTM, ISO)	Interoperable standards for airworthiness, HMI, cybersecurity, vertiports, UTM, and connectivity.	ED-3xx VTOL series; RTCA-EUROCAE joint standards; ASTM vertiports; ISO 5491 expansion; ACJA telecom metrics.	MoCs among SDOs; regulator recognition as AMC/MoC; industry adoption.	RTCA-EUROCAE MoC (2023); EUROCAE-ETSI (2025); ISO 5491.

7.3 INTERNATIONAL CONSORTIUMS AND HARMONIZATION EFFORTS

Amin Vafadar

The AAM ecosystem is converging on a shared premise: without globally compatible rules, common technical standards, and coordinated rollout, eVTOL services will stall at pilot scale. Over the past three years, a multilayered governance lattice has emerged—from ICAO's global policy scaffolding and EASA's Europe-led regulatory packages to transatlantic FAA-EASA implementation procedures and a web of SDOs (EUROCAE, RTCA, ASTM, and ISO) producing interoperable specifications. In practical terms, this lattice is translating into mutual acceptance of approvals, aligned vertiport requirements, U-space/UTM interoperability, and telecom-aviation coordination—the bedrock for cross-border AAM operations (EASA, 2025; ICAO, 2025).

7.3.1 ICAO's Global Architecture

At the apex, ICAO created the Advanced Air Mobility-Study Group (AAM-SG) to craft a harmonized operational and regulatory framework, route detailed tasks to relevant panels, and feed standards/recommended practices (SARPs) into Annexes (ICAO, 2025). The AAM-SG's mandate—formally noted at Assembly sessions—explicitly calls for globally harmonized approaches to unmanned and AAM integration, including interfaces with UTM/ATM and cross-border licensing and operations (ICAO, 2025). In parallel, ICAO's UTM Framework (Edition 4) sets out common services, roles (e.g., U-space/USS), and data exchanges needed for scalable low-altitude operations—a basis on which States can build compatible implementations (ICAO, 2023a; 2025).

7.3.2 Transatlantic Alignment: FAA-EASA TIP Rev. 7 (+ Amendment 1)

Beneath ICAO, the FAA-EASA Technical Implementation Procedures (TIP) are the crucial bilateral "plumbing" for concurrent approvals. Revision 7 (effective Nov. 18, 2023) and Amendment 1 (effective June 10, 2025) update how the agencies accept each other's findings for design approval, production, and environmental certification—streamlining concurrent eVTOL certification/validation and reducing duplicative testing (FAA/EASA, 2023, 2025). The FAA has also signalled deeper policy alignment on eVTOL certification and operations, framing a pathway that complements EASA's work and facilitates mutual recognition over time (FAA, 2024c, 2024d; FAA/EASA, 2023, 2025).

7.3.3 European Leadership: EASA's Innovative Air Mobility (IAM) Package and U-Space

EASA has stepped up with a dedicated IAM regulatory package (July 2025) to enable piloted VTOL/VCA operations within the ATM environment and to give AMC/GM guidance for competent authorities, manufacturers, and operators—codifying an operations-first civil pathway in Europe (EASA, 2025). EASA's materials explicitly reference EUROCAE recorder standards (ED-112/ED-155) and align with U-space/UTM constructs, ensuring that aircraft, operations, and digital services evolve coherently. Complementing this, the EU's U-space concept (via SESAR's CORUS-XUAM ConOps) provides a pan-European services stack (U1-U4) for dense low-altitude operations and has been validated through multi-state demonstrations to prove interoperability and airport/ATM integration (SESAR JU, 2023) (EASA, 2025; SESAR, 2023).

7.3.4 Standards-Development Organizations (SDOs): EUROCAE, RTCA, ASTM, ISO (and Partners)

EUROCAE WG-112 (VTOL) is the anchor for Europe's VTOL standardization, publishing a growing corpus that OEMs and regulators are already referencing. Notable deliverables include ED-300 (VTOL FHAs/PASA methodology), ED-306 (protection from uncontained high-energy fragments/sustained imbalance), ED-305 (information security for VTOL and collaborative systems), and ED-334 (common-mode analysis for lift-thrust systems, enhanced category)—with updates and extensions slated through 2026 (EUROCAE, 2022a, 2025a). These maps directly onto airworthiness, safety, and cybersecurity needs of electric propulsion architectures (EUROCAE, 2022b, 2025a, 2025b).

To avoid fragmented standards, RTCA and EUROCAE renewed their Memorandum of Cooperation (2023), committing to joint, interoperable standards development (e.g., CNS/ATM integration, avionics, and safety) that EASA/FAA can co-reference; EUROCAE also reconfirmed cooperation with ETSI (2025) to align aviation and ICT standardization for connected AAM (RTCA-EUROCAE, 2023; EUROCAE-ETSI, 2025) (RTCA, 2023; EUROCAE, 2025a).

On infrastructure, ASTM F3423/F3423M-22 provides a civil vertiport specification, while the FAA's Engineering Brief 105 (and 105A, 2024) offers federal design guidance; taken together, they are converging toward a shared baseline used by U.S. States and international stakeholders (ASTM, 2022; FAA, 2022; FAA, 2024d). Meanwhile, ISO 5491:2023 (cargo UAS vertiports) has established an international baseline, and the committee is now developing ISO/CD 5491 to broaden the scope and align docking/interface standards across the drone/eVTOL ecosystem (ISO, 2023, 2025) (ASTM, 2022; FAA, 2022, 2024c; ISO, 2023, 2025).

7.3.5 Risk Assessment and Operational Harmonization: JARUS SORA

JARUS provides the de facto Specific Operations Risk Assessment (SORA) methodology used by many authorities (including EASA's specific category) to authorize complex drone/AAM operations. The v2.5 updates refine ground-risk and air-risk models and have been widely adopted—accelerating cross-recognition of operational approvals and providing a template that can evolve toward VTOL passenger missions as automation scales (JARUS/EASA, 2023; JARUS/EASA, 2023).

7.3.6 Telecom-Aviation Convergence: GUTMA-GSMA (ACJA)

Because AAM depends on resilient digital links (C2/C3, surveillance, and data), telecom and aviation SDOs are now co-producing methods and metrics. The Aerial Connectivity Joint Activity (ACJA)—a collaboration between GUTMA and the GSMA—publishes reference methods for cellular link performance and aligns contributions into aviation SDOs (e.g., ASTM Remote ID, and UTM specifications), reducing the risk of incompatibilities between aviation standards and 3GPP-based networks (GSMA, 2023).

- **Synthesis: what "harmonization" now means in practice:** By mid-2025, harmonization is no longer aspirational; it is operational. ICAO is defining the common lexicon and baseline frameworks; EASA has issued Europe-wide IAM rules that integrate with U-space; FAA-EASA TIP Rev. 7 (Amendment 1) enables concurrent certification; and SDOs are publishing mutually recognized standards spanning airworthiness, safety, cybersecurity, UTM, and vertiports (Table 7.2). The net effect is a pathway whereby an eVTOL design certified in one jurisdiction can move through validation elsewhere with fewer deltas; a vertiport built to ASTM/FAA/ISO guidance can support multiple OEMs; and digital services—UTM and cellular—operate to common performance expectations. This is precisely the precondition for cross-border AAM that the sector has sought (EASA, 2025; ICAO, 2025).

7.4 CROSS-BORDER TRAINING AND CERTIFICATION PROGRAMS

Amin Vafadar

Building a mobile, internationally deployable AAM workforce requires portable qualifications for flight crew and remote pilots, mutual recognition mechanisms between authorities, and aligned training philosophies (competency-based and evidence-based). Over the last two years, regulators have begun to harden these elements into rules, guidance, and bilateral procedures—with ICAO setting the global pillars, EASA leading within Europe, and the FAA establishing a powered-lift licensing framework that can be bridged internationally.

- **Global baselines (ICAO):** ICAO's *Annex 1—Personnel Licensing* now provides the umbrella minimum standards for licenses and ratings, including provisions for Remote Pilot License (RPL) in the RPAS context; training guidance is elaborated in *PANS-TRG (Doc 9868)*, which formalizes competency-based training and assessment (CBTA) across personnel categories. For AAM, these texts are the logical reference points for harmonizing both on-board eVTOL pilot training (e.g., CBTA principles applied to transition/energy management and multimode operations) and remote-pilot or remote-supervisor roles as automation scales. States then transpose Annex 1 minima into national rules, enabling cross-recognition of licenses and credits under existing bilateral/multilateral arrangements.
- **United States: powered-lift licensing and the SFAR bridge:** In October 2024, the FAA adopted a final rule that both amends long-standing parts

TABLE 7.2
Harmonization layers and instruments

Layer/Institution	Instrument (Latest)	Scope	2024–2025 Status	Cross-border Value
ICAO	AAM-SG work program; UTM Framework Ed.4	Global lexicon; high-level ops frameworks; UTM services & roles.	Active SG; frameworks referenced by States.	Common language for States; baseline for mutual recognition.
FAA-EASA	TIP Rev.7 + Amendment 1 (2025)	Design/production/env. acceptance procedures.	Updated acceptance flows for new tech; supports concurrency.	Reduces duplicate tests; accelerates validations.
EASA (EU)	Innovative Air Mobility (IAM) package; U-space rules/AMC-GM	Piloted VTOL ops within ATM; digital services via U-space.	Issued July 2025; implementation by MS underway.	Replicable EU template for ops & services.
SDOs (EUROCAE/ RTCA)	ED-300/305/306/334; joint MoC	VTOL safety, info-security, fragment protection, common-mode analysis.	New/updated EDs through 2025–2026; joint development model.	Interoperable standards regulators can reference.
ASTM/FAA/ISO	F3423 vertiports; EB-105/105A; ISO 5491	Civil vertiport design; federal guidance; international vertiport spec.	Converging baseline used by airports & cities.	Supports multi-OEM ops; portable infrastructure approvals.
JARUS	SORA v2.5	Risk assessment for complex ops (UAS/AAM).	Widely adopted by authorities; evolving toward VTOL CAT use.	Cross-recognition of operational approvals.
GUTMA-GSMA (ACJA)	Connectivity metrics/ methods	Cellular performance for aviation (C2/C3, RID, data).	Methods referenced in sector guidance.	Aligns telecom & aviation expectations for AAM.

and issues a 10-year Special Federal Aviation Regulation (SFAR) to stand up the first cohorts of powered-lift pilots and instructors, clarify operating rules, and integrate powered-lift into the NAS. This rule provides the immediate regulatory pathway for eVTOL pilot certification and training program approval in the United States, and it explicitly frames a transition to enduring rules as the industry matures. These provisions are already accompanied by agency FAQs and implementation materials for operators and training organizations.

- **Europe: EASA's operations-first pathway and training enablers:** In July 2025, EASA issued its IAM regulatory package to enable piloted VTOL/VCA operations in the ATM environment, with associated AMC/GM for authorities, manufacturers, and operators. While IAM centers on operations, EASA has, in parallel, set up processes to qualify new training technologies—notably Special Conditions for Flight Simulation Training Devices (FSTD) that cover virtual reality and novel aircraft types (including eVTOL). These mechanisms allow training credits to be earned on devices representing new flight phases (e.g., conversion/transition) before fleets are large, and they reduce cross-border friction when devices are accepted by multiple authorities.
- **Mutual recognition: transatlantic conversion and concurrent approvals:** For decades, the U.S.-EU BASA has included Annex 3 (Pilot Licensing) and FAA AC 61–143, which together describe how an FAA certificate can be converted to an EU Part-FCL license (and vice versa), what knowledge/skill gaps must be bridged, and how currency is credited—machinery that AAM can leverage for eVTOL pilots as powered-lift/VTOL ratings propagate. In parallel, the FAA-EASA TIP—while focused on design/production—are being updated to streamline concurrent approvals around the eVTOL ecosystem, which indirectly supports aligned training syllabi and device acceptances. Practically, these instruments let an operator develop one core curriculum and bridge modules to satisfy both systems, reducing duplicate training time while preserving safety intent.
- **Cross-border initiatives and regional partnerships:** Authorities are also using MoUs to expedite alignment where traffic will be international from the outset. A notable example is the EASA-CAAS (Singapore) MoU (Oct. 2022), which explicitly targets common safety standards for VTOL operations, the competencies of personnel, and requirements for training organizations, including hot/humid operations—an archetype for EU-ASEAN bridging of syllabi and approvals. Within Europe, EASA's IAM dovetails with U-space implementation and pan-EU drone rules, giving a continental template onto which national AAM training policies can be mapped with minimal divergence.
- **Training devices and dual-qualification proof points:** Cross-recognition is accelerating at the device level. EASA has established a dedicated Special Conditions pathway for new FSTD technologies, and in 2024, Loft Dynamics announced the first VR simulator with both EASA and FAA qualification, demonstrating that bi-national training credit on novel devices

is attainable when fidelity and evaluation methods are aligned. Such precedents are instructive for forthcoming eVTOL FSTD qualifications and for distributed, cross-border training networks that rely on the same high-fidelity devices in multiple States.

- **Implications for operators and ATOs:** The converging picture is a two-layer system: (i) global CBTA principles and license minima (ICAO) that anchor safety and portability and (ii) regional/national licensing/rating constructs (EASA IAM and FAA powered-lift SFAR) that are bridged via BASA conversion and targeted MoUs. For airlines, helicopter incumbents, and new-entrant ATOs planning multi-city networks (EU, Middle East, and ASEAN), this means designing modular syllabi: a common CBTA core (vehicle/energy management, mode transitions, abnormal procedures, and U-space/UTM interfaces), plus jurisdiction-specific bridge modules (local air law, radio/phraseology, and operational approvals). Where feasible, dual-qualified FSTDs should be used to front-load training and standardize competencies. As remote supervision grows, the same logic extends to RPL and remote-pilot roles, using Annex 1 and PANS-TRG as the global touchstones and regional rules for the operational specifics. Table 7.3 summarizes the global training.

7.5 COMMUNITY ENGAGEMENT AND PUBLIC POLICY DEVELOPMENT

Amin Vafadar

Securing a durable "social licence to operate" is a precondition for any scaled AAM deployment. European evidence shows that citizens express an initially positive attitude toward UAM but condition acceptance on guarantees around safety, noise, security, and environmental protection; noise in particular emerges as the second most-cited concern after safety (EASA, 2021). Accordingly, community engagement and public policy must proceed in lockstep with certification and infrastructure rollout, rather than as an afterthought (FAA, 2023a; WEF, 2025).

A robust policy architecture for engagement is now visible at three layers. At the global/regional level, regulators frame expectations for early public involvement and impact mitigation. The FAA's *I28* plan explicitly requires engagement beyond airspace changes—placing community, environmental, and infrastructure issues alongside certification milestones (FAA, 2023b). In Europe, EASA's 2025 IAM package embeds an operations-first pathway that must be reconciled with local land use, energy, and vertiport availability—implicitly demanding upstream coordination with municipalities (EASA, 2025). At the national/local level, aviation and city authorities are issuing practical guidance. The UK CAA's CAP 2538 advises that vertiports intended for public transport may need aerodrome licensing and must be integrated into existing planning and environmental regimes, while CAP 3041 provides structured methods for airport-community environmental dialogue that can be adapted for AAM. At the city network level, the EU's UIC2 ("Urban Air Mobility Initiative Cities Community") and associated Smart Cities resources codify multi-level governance and SUMP-based integration to align UAM with broader urban mobility goals (UIC2/Smart Cities Marketplace, 2022).

TABLE 7.3
Cross-border training, licensing, and device acceptance

Dimension	Europe (EASA)	United States (FAA)	Global/Bilateral Mechanism	Implications for Operators/ATOs
License/rating construct	VTOL type rating added to CPL/ATPL (Part-FCL); IAM enables piloted VTOL ops.	Powered-lift category & ratings (SFAR No. 120).	ICAO Annex 1 minima; BASA Annex 3 conversion.	Design modular syllabi that map both constructs.
Training philosophy	CBTA/EBT embedded; device credit via Special Conditions for FSTDs.	Outcome-focused within a prescriptive framework; simulator credits allowed.	ICAO Doc 9868 CBTA; national AMC/AC.	Tailor time in device vs. aircraft; assess competencies continuously.
FSTD/devices	SC process for novel FSTDs (incl. VR); dual acceptance emerging.	Qualification pathways; early acceptance for powered-lift sims.	Bi-national device qualification precedents (e.g., VR).	Use dual-qualified devices to standardize skills across regions.
Ab initio & transition	Near-term focus on professional pilots; ab initio signalled later.	Ab initio allowed for powered-lift; practical scaling via sims.	MoUs and guidance to bridge gaps.	Blend transition modules (heli ↔ fixed-wing) with energy/mode management.
Mutual recognition	Part-FCL ↔ FAA conversion supported via BASA/AC 61-143.	As left.	BASA Annex 3; TIP context.	Reduce duplicate training with bridge modules & standardized records.
Remote pilot/supervisor	EU drone rules + U-space context; pathways for future remote supervision.	Emerging for advanced entrants; RID/UTM alignment.	ICAO Annex 1 + PANS-TRG anchors.	Plan curricula for future remote roles alongside piloted ops.

Noise policy and measurement have moved from concept to concrete instruments. EASA has published guidance and communications on noise measurement for drones (≤600 kg) to create a consistent reporting method and public repository and has since signalled the development of eVTOL-specific noise standards and limits to underpin citizen acceptance (EASA 2025). In parallel, several authorities and expert groups recommend multipronged mitigation: vertiport siting away from sensitive receptors, corridor and procedure design for minimum tonal/impulsive components, operational curfews, and transparent publication of noise footprints (UIC2, 2022). Notably, empirical cases show that iterating technology in response to community feedback works: Wing's Canberra pilot prompted a redesign of propeller blades that reduced overall loudness and altered tonal characteristics to address annoyance, alongside route randomization to distribute exposure (RSG, 2021).

Furthermore, the engagement practice is shifting from one-way "informing" to co-production. Best-practice toolkits now emphasize (i) early scoping with residents before site selection; (ii) participatory vertiport siting using mapped noise contours and alternative scenarios; (iii) open data portals showing real-time flight activity and environmental metrics; and (iv) benefit-sharing mechanisms (e.g., integrating AAM with public transport, setting equity targets, or committing community funds near vertiports). The FAA's I28 process references these broader engagement domains, and NASA/FAA convened multiagency community workshops to familiarize local officials and residents with AAM use cases and constraints. On the European side, UIC2 and SUMP guidance recommend embedding AAM in existing mobility plans, treating aerial modes as complementary rather than exceptional. The World Economic Forum's 2025 framework adds 11 responsible implementation principles—linking safety, environmental stewardship, transparency, and inclusivity to social acceptance targets—providing a common language for public-private commitments (WEF, 2025).

Additionally, crucially, policy is becoming evidence-led. The EASA acceptance study recommends that authorities "take measures to meet citizens' expectations" and publish clear guarantees on safety, security, noise, and environment; it also notes higher support for common-interest missions (e.g., medical/emergency transport), which many cities now prioritize as first deployments to build trust (EASA, 2021). Contemporary public-health reviews by the UK CAA synthesize emerging evidence on human response to novel aircraft noise, enabling proportionate limits and adaptive rules as psychoacoustic understanding matures. Together, these instruments support an adaptive governance model in which operators earn and sustain social license through demonstrable performance, transparent data, and responsiveness to feedback.

- **Policy takeaway:** For civil AAM, engagement and policy must be designed as core system elements, not mitigation layers. The minimal viable set includes: a publicly documented engagement plan; measurable community KPIs (noise, complaints, and equity of access); open data commitments; conformance with national vertiport/land use guidance; and alignment with international frameworks so that practices are portable across borders. Where this package is in place—and visibly so—AAM earns legitimacy to scale (Table 7.4).

TABLE 7.4
Community engagement and public-policy architecture (from principles to metrics)

Governance Layer	Objectives	Core Instruments	Typical Measures	Exemplars/Sources	KPIs to Track
Global/regional regulators	Set expectations for early engagement; align safety/noise frameworks.	Innovate28; EASA IAM; UIC2/SUMP guidance; WEF responsible principles.	Public consultations; phased rollouts; transparency requirements.	FAA (2023a); EASA (2025); WEF (2025).	Noise contours published; complaints trend; equity of access.
National aviation + city planning	Integrate vertiports into planning/env. regimes; protect communities.	CAP 2538; CAP 3041; national land use/energy rules.	Aerodrome licensing where applicable; curfews; routing away from sensitive receptors.	UK CAA guidance (2023–2024).	Permit timelines; compliance findings; mitigation closure rate.
City/metropolitan networks	Embed AAM within mobility plans; coordinate with transit.	UIC2; SUMP toolkits; open data portals.	Participatory siting; benefit-sharing; real-time dashboards.	UIC2/Smart Cities Marketplace.	Modal shift; PT integration usage; community sentiment index.
Project-level (operators)	Earn social license; monitor & mitigate impacts; be transparent.	Engagement plans; environmental statements; ops restrictions.	Noise footprints; route randomization; incident reporting.	Wing Canberra case; FAA community workshops.	Event rates; noise exceedances; response time to complaints.

7.6 CASE STUDY: MULTIREGIONAL STAKEHOLDER ENGAGEMENT IN EUROPE, THE MIDDLE EAST, AND ASEAN

Amin Vafadar

The selected case study examines three credible, civil deployments that illustrate how stakeholders align policy, infrastructure, standards, and operations to accelerate AAM: (i) Europe—Rome/Paris testbeds under EASA's orbit; (ii) Middle East—Dubai's government-anchored launch program; and (iii) ASEAN—a Singapore-Malaysia cross-border logistics corridor and Singapore's long-running vertiport/test-flight ecosystem. Together they demonstrate complementary governance models—EU "regulatory-first" with U-space/UTM proofs, a Gulf "concessionary" model with a public authority as orchestrator, and an ASEAN "corridor" model that begins with uncrewed logistics and builds toward passenger services.

- **Europe: Rome (UrbanV/Fiumicino) and Paris (Pontoise) as living testbeds:** In October 2022, Aeroporti di Roma, UrbanV, Volocopter, and partners inaugurated Italy's first AAM testing vertiport at Rome Fiumicino and conducted the country's first crewed public eVTOL test flights, establishing an EU-compliant infrastructure and operations sandbox. UrbanV has continued to use the Rome testbed to evaluate operational technologies—for example, AI bird-strike monitoring—that feed into EASA-aligned safety cases and future standardization (Aeroporti di Roma/UrbanV and AIN Online). These activities align with the EU's U-space trajectory (SESAR/Eurocontrol) and illustrate how airport operators, OEMs, and regulators can iterate on ground procedures, energy layouts, and safety mitigations before scaling to commercial service (UrbanV and SESAR JU) (Aeroporti di Roma, 2022; AIN Online, 2024; SESAR JU, 2023).

In Paris, a parallel coalition—Groupe ADP, RATP Group, Choose Paris Region, and industry partners—stood up Europe's first fully integrated vertiport terminal/testbed at Pontoise-Cormeilles and ran iterative demonstrations with Volocopter in the run-up to the 2024 Games (Groupe ADP press; Asian Sky Group; and Electronomous). Although full passenger operations were not certified for the Olympics, the operational flight testing campaign progressed—validating ground flows, passenger processes, and U-space interfaces that EASA can reference in guidance and cities can adopt in planning (Aerospace Testing International; AP). The Paris and Rome efforts, together with CORUS-XUAM demonstrations, show how transnational EU initiatives convert standards into practice and derisk early operations through airport-anchored governance (AP, 2024; Asian Sky Group, 2022; Groupe ADP, 2021; SESAR JU, 2023).

To summarize, the EU model ties living-lab operations to EASA's IAM policy and U-space service stacks, enabling replicable procedures and infrastructure templates across Member States. Airport operators (ADP and ADR) act as "ecosystem integrators," convening OEMs, vertiport developers, ANSPs, and cities, while SESAR projects supply the ConOps and data needed for harmonization.

- **Middle East: Dubai's concessionary model under a single public orchestrator:** Dubai's RTA has executed a definitive agreement with Joby

Aviation and Skyports Infrastructure to launch an air-taxi network with four initial vertiports (DXB, Palm Jumeirah, Downtown, and Marina) and six-year exclusive operating rights for Joby—anchored to a 2025–2026 service window (Joby IR; Skyports). In late 2024, Joby and RTA announced work commencing on the first DXB vertiport, and by mid-2025, Joby reported delivery and local testing of its first aircraft in the Emirate; Reuters and other outlets documented early demonstration flights and reiterated the concession's scope (Joby; The Verge; and Reuters). The governance configuration is notable: RTA orchestrates policy, sites, and multimodal integration; the GCAA/DCAA and Dubai ANSPs supervise airworthiness/operations and airspace procedures; and Skyports delivers the vertiport network—offering a clear, bankable pathway for private capital while keeping public objectives (safety, connectivity, and noise) central (Joby, 2024a, 2024b, 2025; Reuters, 2025; Skyports, 2024; The Verge, 2025).

The Dubai model shows how a single public authority can compress timelines by aligning approvals, infrastructure, and commercial rights—reducing counterparty risk for OEM-operators and providing visible milestones (site starts, aircraft delivery, and route setup) that build public and investor confidence.

- **ASEAN: a cross-border drone-delivery corridor and Singapore's vertiport lineage:** In February 2024, Aerodyne (Malaysia) and DroneDash (Singapore) announced a shore-to-shore cross-border delivery corridor linking Singapore with Johor Bahru/Iskandar Malaysia, coordinated with both countries' regulators. The plan calls for dual-city 5G roaming, sat-nav augmentation, and aircraft in the ≤30 kg/150 km h^{-1} class to move urgent goods—an archetype of bilateral UTM/telecom/airspace integration for uncrewed AAM (Unmanned Airspace; TechWire Asia) (Unmanned Airspace, 2024; TechWire Asia, 2024) (Unmanned Airspace, TechWire Asia).

This builds on Singapore's decade-long, regulator-backed ecosystem: CAAS, EDB, JTC/Seletar Aerospace Park, Skyports, and Volocopter have collaborated since 2019, when Volocopter flew a manned public test over Marina Bay and showcased the world's first full-scale vertiport prototype (VoloPort) with Skyports. Subsequent MOUs laid the groundwork for vertiport facilities and MRO/training at Seletar, with roadmaps to multiple vertiports and eventual cross-border passenger links (EDB; Skyports; Aviation Week; and FlightGlobal). The EASA-CAAS MoU (2022) explicitly targets alignment on VTOL safety standards and personnel competencies, reinforcing international portability of training/approvals (EASA press and eVTOL Insights) (Aviation Week, 2022; FlightGlobal, 2022; Volocopter/EDB, 2019–2022).

ASEAN's corridor-first approach shows how uncrewed services can catalyze cross-border harmonization (permits, UTM, and telecom QoS), while Singapore's regulator-led prototyping provides a repeatable template for vertiport siting, passenger processing, and stakeholder outreach—positioning the region for passenger eVTOL services as certification pathways mature.

- **Cross-regional lessons:** Based on the presented theory and the presented case studies, we have to focus on the following cross-regional lessons:

1. A public integrator accelerates progress. The EU model elevates airport operators and SESAR programs as integrators; Dubai relies on RTA; Singapore coordinates through CAAS/EDB/JTC. Each model formalizes roles and timelines and makes accountability legible (Groupe ADP; Joby/Skyports; EDB/CAAS).
2. Standards and demos must co-evolve. Rome/Paris tests fed EASA/U-space ConOps; Dubai's staged demos are tied to site builds and operating concessions; ASEAN's corridor couples UTM and 5G performance metrics to regulatory approvals (SESAR; Reuters; and Unmanned Airspace).
3. Measurable milestones build trust. Concrete, public milestones—first vertiport works, first local aircraft delivery/demos, first cross-border flights—enable transparent progress reviews and inform community engagement and investor due diligence (Joby; The Verge; and UrbanV).

7.7 CONCLUSION

Dimitrios Ziakkas, Amin Vafadar, and Debra Henneberry

Human performance is the through-line that connects certification, training, cockpit design, and operational oversight in eVTOL services. By mid-decade, AAM has transitioned from an engineering experiment to an institutional project whose success will be judged less by the novelty of the aircraft and more by the reliability of the system that surrounds it. Three conclusions stand out. First, durable progress is coalition-built. Where airport operators, OEMs, service providers, municipalities, and regulators sign onto a shared playbook—articulating roles, timelines, and performance duties—deployment accelerates without compromising safeguards. The European airport-anchored testbeds and Dubai's concessionary model represent different governance choices that nevertheless converge on the same discipline: explicit milestones, auditable procedures, and an allocation of risk that capital providers, residents, and authorities can understand.

Second, harmonization has matured from aspiration to operating practice. ICAO's policy scaffolding, the FAA-EASA TIP, and the standards ecosystem now enable concurrent approvals, interoperable vertiports, and common expectations for digital services. This reduces duplication in evidence generation, allows operators to reuse core syllabi with jurisdictional bridge modules, and gives cities confidence that infrastructure built once will support multiple OEMs. It also means that human-factors guidance—on handling qualities, HMI, and training assessment—travels with the aircraft, allowing safety assurances to be portable rather than bespoke.

Third, social license is earned through performance and openness. Communities accept new aviation when they can see and verify the controls that matter to them: noise and exposure footprints, hours of operation, environmental baselines, and fair access. The most successful programs have shifted from one-way announcements to co-produced siting and transparent data sharing, treating predictability and accountability as primary safety features. In practice, this entails measurable community

KPIs, open portals for flight and environmental data, and a willingness to iterate technology and procedures in response to feedback.

Taken together, these dynamics suggest a pragmatic path forward. Begin where governance capacity is strongest; build to common standards; choose corridor and network designs that privilege predictability; and publish evidence—on safety, environment, and service quality—that withstands independent scrutiny. When AAM is framed and managed as a human-centered public service rather than a technology showcase, it gains the legitimacy needed to scale from a cluster of demonstrations to a dependable layer of urban and regional mobility.

REFERENCES

Aeroporti di Roma. (2022). *Italy's first vertiport deployed at Fiumicino Airport; first crewed eVTOL public testflights*. https://www.adr.it/en/web/aeroporti-di-roma-en/viewer?_3_WAR_newsportlet_jspPage=%2Fhtml%2Fnews%2Fdetails.jsp&_3_WAR_newsportlet_nid=26001173&_3_WAR_newsportlet_redirect=%2F

AIN Online. (2024, April 5). *Rome's vertiport tests AI-powered bird monitoring system*. https://www.ainonline.com/aviation-news/futureflight/2024-04-05/romes-vertiport-tests-ai-powered-bird-monitoring-system

American Press (AP). (2024, August 11). *Air taxis failed to get certified for the Paris Olympics* https://apnews.com/article/olympics-2024-paris-flying-taxis-82af3b21726dc6ebbc99da076b98d14b

Asian Sky Group. (2022). *Europe's first vertiport testbed launched in Paris*. https://www.asianskygroup.com/europes-first-vertiport-testbed-launched-in-paris-886/

ASTM International. (2022). *F3423/F3423M-22: Standard specification for vertiport design*. https://www.astm.org/f3423_f3423m-22.html

Aviation Week. (2022, February 15). *Volocopter unveils major Singapore eVTOL push*. https://aviationweek.com/shownews/singapore-airshow/volocopter-unveils-major-singapore-evtol-push

Delta Air Lines. (2022). *Delta and Joby partner to pioneer home-to-airport transportation*. https://news.delta.com/delta-joby-aviation-partner-pioneer-home-airport-transportation-customers

EASA/Civil Aviation Authority of Singapore (CAAS). (2022, October 18). *Memorandum of understanding on Urban Air Mobility*. https://www.easa.europa.eu/en/newsroom-and-events/press-releases/caas-and-easa-sign-memorandum-understanding-urban-air-mobility

EUROCAE & European Telecommunications Standards Institute (ETSI). (2025, May 22). *Updated memorandum on ICT and aviation standardisation*. https://www.eurocae.net/eurocae-and-etsi-reconfirm-cooperation-with-updated-memorandum-of-understanding-on-ict-and-aviation-standardisation/

Eurocontrol. (2022). *U-space services implementation monitoring report (Edition 2022)*. https://www.eurocontrol.int/sites/default/files/2022-11/eurocontrol-u-space-services-implementation-monitoring-report-2022.pdf

Eurocontrol. (2024). *European network of U-space demonstrators/stakeholders*. https://www.eurocontrol.int/product/european-network-u-space-demonstrators

European Organisation for Civil Aviation Equipment (EUROCAE). (2022a). *ED-300: Guidance on conducting an AFHA and PASA for VTOL*. https://www.eurocae.net/working-group/wg-112-sg-3/

European Organisation for Civil Aviation Equipment (EUROCAE). (2022b). *ED-306: VTOL protection from uncontained high-energy fragments*. https://www.eurocae.net/working-group/wg-112-sg-2/

European Organisation for Civil Aviation Equipment (EUROCAE). (2025a). *ED-305: Information security guidance for VTOL and collaborative systems*. https://www.eurocae.net/working-group/wg-112-sg-3/

European Organisation for Civil Aviation Equipment (EUROCAE). (2025b). *ED-334: Common-mode analysis for lift–thrust systems (enhanced category)*. https://www.eurocae.net/working-group/wg-112-sg-2/

European Union Aviation Safety Agency (EASA). (2021). *Study on the societal acceptance of Urban Air Mobility in Europe*. https://www.easa.europa.eu/sites/default/files/dfu/uam-full-report.pdf

European Union Aviation Safety Agency (EASA). (2025, July 9). *EASA steps up regulatory framework for Innovative Air Mobility*. https://www.easa.europa.eu/en/newsroom-and-events/news/easa-steps-regulatory-framework-innovative-air-mobility

Federal Aviation Administration (FAA). (2022, September 26). *FAA releases vertiport design standards to support safe integration of AAM (EB-105)*. https://www.faa.gov/newsroom/faa-releases-vertiport-design-standards-support-safe-integration-advanced-air-mobility

Federal Aviation Administration (FAA). (2023a). *Advanced Air Mobility (AAM) implementation plan—Innovate28*. https://www.faa.gov/sites/faa.gov/files/AAM-I28-Implementation-Plan.pdf

Federal Aviation Administration (FAA). (2023b, July 18). *FAA issues implementation plan outlining steps to usher in AAM*. https://www.faa.gov/newsroom/faa-issues-implementation-plan-outlining-steps-usher-advanced-air-mobility

Federal Aviation Administration (FAA). (2024a). *Powered-lift SFAR—Frequently asked questions*. https://www.faa.gov/air-taxis/FAQ

Federal Aviation Administration (FAA). (2024b, June 10). *Statement on eVTOL aircraft certification*. https://www.faa.gov/newsroom/faa-statement-evtol-aircraft-certification

Federal Aviation Administration (FAA). (2024c, October 22). *Integration of powered-lift: Pilot certification and operations; miscellaneous amendments (Final Rule & 10-year SFAR)*. https://www.faa.gov/newsroom/integration-powered-lift-pilot-certification-and-operations-miscellaneous-amendments

Federal Aviation Administration (FAA). (2024d, December 27). *Engineering brief 105A: Vertiport design*. https://www.faa.gov/airports/engineering/engineering_briefs/eb_105a_vertiports

Federal Aviation Administration (FAA) & European Union Aviation Safety Agency (EASA). (2023/2025). *Technical implementation procedures—Revision 7 and amendment 1 (effective June 10, 2025)*. https://www.faa.gov/aircraft/air_cert/international/bilateral_agreements/eu/tip

FlightGlobal. (2022, February 14). *Singapore explores developing UAM hub at Seletar*. https://www.flightglobal.com/singapore-2022/singapore-explores-developing-uam-hub-at-seletar-aerospace-park/147518.article

Groupe ADP. (2021, November 25). *Pontoise airfield: First UAM trials launched*. https://presse.groupeadp.fr/pontoiseairfield-uamtests/

GSMA. (2023). *North America drones white paper (ACJA overview)*. https://www.gsma.com/about-us/regions/north-america/wp-content/uploads/2023/12/GSMA-North-America-Drones-White-Paper_Final-Nov_30_2023.pdf

International Civil Aviation Organization (ICAO). (2023a). *Annex 1—Personnel licensing* (incl. RPAS/RPL provisions).

International Civil Aviation Organization (ICAO). (2023b). *Procedures for air navigation services—training* (Doc 9868, 2nd ed.).

International Civil Aviation Organization (ICAO). (2023c). *UTM framework (Edition 4)*. https://www.icao.int/sites/default/files/left-menu-pdfs/UTM%20Framework%20Edition%204.pdf

International Civil Aviation Organization (ICAO). (2025). *State of global aviation safety 2025* (AAM-SG reference). https://www.icao.int/sites/default/files/sp-files/safety/Documents/ICAO_SR_2025.pdf

International Organization for Standardization (ISO). (2023). *ISO 5491:2023—Vertiports (cargo UAS)*. https://www.iso.org/standard/81313.html

International Organization for Standardization (ISO). (2025). *ISO/CD 5491—Vertiports (broader scope, under development)*. https://www.iso.org/standard/89706.html

JARUS/European Union Aviation Safety Agency (EASA). (2023). *SORA v2.5 updates (workshop summary)*. https://www.easa.europa.eu/sites/default/files/dfu/sora_workshop_feb_2023_-_4_summary_of_changes_sora.pdf

Joby Aviation. (2024a, February 11). *Definitive agreement with Dubai RTA to launch air-taxi service*. https://ir.jobyaviation.com/news-events/press-releases/detail/87/joby-to-launch-air-taxi-service-in-uae

Joby Aviation. (2024b, November 12). *Beginning of work on first Dubai vertiport (DXB)*. https://ir.jobyaviation.com/news-events/press-releases/detail/115/joby-announces-beginning-of-work-on-first-dubai-vertiport

Joby Aviation. (2025, June 30). *Cements global lead with Dubai programme (six-year exclusivity)*. https://ir.jobyaviation.com/news-events/press-releases/detail/132/joby-cements-global-lead-in-air-taxi-industry-with-dubai

Loft Dynamics. (2024, July 31). *World's first VR FSTD with FAA qualification; also EASA-qualified*. https://www.loftdynamics.com/loft-dynamics-becomes-worlds-first-vr-flight-simulation-training-device-to-receive-faa-qualification/

Reuters. (2025, July 1). *Dubai aims to beat traffic with 2026 Joby air-taxi liftoff*. https://www.reuters.com/world/middle-east/dubai-aims-beat-traffic-with-2026-joby-air-taxi-liftoff-2025-07-01/

Roland Berger. (2018). *Urban Air Mobility: The rise of a new mode of transportation*. https://www.rolandberger.com/publications/publication_pdf/Roland_Berger_Urban_Air_Mobility.pdf

Roland Berger. (2020). *Urban Air Mobility (update)*. https://www.rolandberger.com/publications/publication_pdf/roland_berger_urban_air_mobility_1.pdf

RSG. (2021). *Three considerations around drone noise (Wing Canberra case)*. https://rsginc.com/wp-content/uploads/2021/12/RSG_White-Paper_Drone-Noise_Updated-12-2021.pdf

RTCA. (2023, April 26). *RTCA and EUROCAE renew memorandum on joint standards development*. https://www.rtca.org/news/advancing-aviation-standards-rtca-and-eurocae-renew-commitment-through-updated-memorandum-of-cooperation/

SESAR Joint Undertaking. (2023). *U-space concept of operations (4th ed.)/CORUS-XUAM*. https://www.sesarju.eu/sites/default/files/documents/reports/U-space%20CONOPS%204th%20edition.pdf

Skyports Infrastructure. (2024, February 11). *Skyports, RTA and Joby to launch air-taxi service in Dubai*. https://skyports.net/skyports-rta-and-joby-to-launch-air-taxi-service-in-dubai/

TechWire Asia. (2024, February). *Can cross-border drone deliveries between Malaysia and Singapore work?* https://techwireasia.com/2024/02/can-cross-border-drone-deliveries-between-malaysia-and-singapore-work/

UIC2/Smart Cities Marketplace. (2022). *UIC2 overview*. https://smart-cities-marketplace.ec.europa.eu/sites/default/files/2022-03/UIC2_Overview_February-2022.pdf

Unmanned Airspace. (2024, February 22). *Aerodyne and DroneDash to launch cross-border deliveries between Malaysia and Singapore*. https://www.unmannedairspace.info/aam-uam-route-and-programme-news/aerodyne-and-dronedash-agree-to-launch-cross-border-deliveries-between-malaysia-and-singapore/

The Verge. (2025, July). *Joby delivers first aircraft to Dubai as air-taxi service nears launch.* https://www.theverge.com/news/694515/joby-delivers-first-aircraft-to-dubai-as-air-taxi-service-nears-launch

World Economic Forum (WEF). (2025). *Advanced Air Mobility: Paving the way to responsible implementation.* https://reports.weforum.org/docs/WEF_Advanced_Air_Mobility_2025.pdf

8 Scaling and Adaptability of AAM–eVTOL Networks

Dimitrios Ziakkas, Debra Henneberry, and Amin Vafadar

8.1 INTRODUCTION

Scaling Advanced Air Mobility (AAM) from isolated demonstrations to everyday service is not a single-technology problem; it's a place-sensitive, system-of-systems transition. The core argument of this chapter is that networks scale credibly only when four layers are advanced together and tailored to local context: (1) fit-for-purpose infrastructure and energy systems (from rooftop pads to megawatt-class hubs), (2) land use and zoning instruments that balance access with community protections, (3) airspace and operational concepts that layer safely from corridor pilots to high-tempo nodes, and (4) demand-led service design that matches passenger and cargo use-cases to urban, suburban, and rural geographies. Regulators have already begun to codify this approach. FAA's *Innovate28* points to "crawl-walk-run" integration at key sites, while EASA's Innovative Air Mobility (IAM) framework embeds operations, landing-site availability, and energy planning into approvals—signalling that ground and air constraints must be co-managed from day one.

The chapter proceeds in four parts. Section 8.2 details how infrastructure, energy provisioning, land use, operations, demand, and phasing strategies differ by setting—and why mixed passenger-cargo portfolios accelerate scale. Section 8.3 shows how regulatory sandboxes, phased approvals, and risk-based methods (e.g., SORA 2.5) let emerging markets move from trials to stable rules without waiting for complete, top-down regulation. Section 8.4 brings the smart-city lens: AAM becomes truly useful when digital platforms (operator/vertiport OS, U-space/UTM, and MaaS) and physical interchanges (airports, rail hubs, and district centers) are designed as one system—often with autonomous shuttles solving the first/last mile. Section 8.5 distills lessons from Southern Europe (Rome-Lazio, Greece, and Spain), where airport-city trunks, tourism-driven inter-island links, and U-space-enabled medical logistics demonstrate transferable blueprints for growth. Throughout, we connect these operational moves to business viability and societal value: earlier chapters on revenue models and ROI emphasize that scaling choices must line up with workable pricing, utilization, and investment cases, and that benefits accrue beyond operator P&L to jobs, productivity, and resilience.

DOI: 10.1201/9781003685555-8

In short, the path to scale is pragmatic and iterative: start where demand and governance capacity are strongest (often airport anchors and public-benefit cargo), instrument operations to produce evidence, and broaden privileges as safety and performance data accumulate. Doing so aligns with the ROI and scenario-forecasting disciplines highlighted earlier and increases the odds that capital, communities, and regulators remain in step as networks densify.

8.2 SCALING eVTOL NETWORKS ACROSS URBAN, SUBURBAN, AND RURAL ENVIRONMENTS

Amin Vafadar, Tomislav Lang, and Dimitrios Ziakkas

The implementation of AAM operations from demonstration to daily service requires a place-sensitive approach: the infrastructure, operating concepts, and policy instruments that work in dense urban cores are not identical to those needed in suburbs or sparsely populated regions. A practical pathway combines (i) fit-for-purpose vertiport infrastructure and energy systems, (ii) land use and zoning integration that protects communities while enabling access, (iii) airspace/operations concepts that can be layered from low to high density, and (iv) demand-led service design spanning both passenger and cargo use cases. This systems view is reflected in the FAA's *Innovate28* implementation plan (phased "crawl-walk-run" integration to one or more key sites by 2028) and in EASA's 2025 IAM package, which embeds operations, pre-flight preparation, availability of landing sites, and energy management into an integrated regulatory fabric (EASA, 2025; FAA, 2023a).

- **Infrastructure and energy: from rooftop pads to megawatt-class nodes:** Urban networks favor retrofitting existing heliports and rooftops into vertiports, supplemented by purpose-built hubs at intermodal nodes (rail stations and airports). FAA's *Engineering Brief 105A* (Dec. 27, 2024) provides the current U.S. baseline for vertiport geometry, obstacle protection, markings/lighting, and landside features, explicitly as supplemental guidance to AC 150/5390-2D (heliports)—a pragmatic bridge that allows cities and airport operators to convert or design sites now and iterate as standards mature (FAA, 2024). In Europe, EASA's IAM framework emphasizes assured availability of landing sites (including diversion locations) and fuel/energy management as part of operations approval, pushing operators to plan ground infrastructure and reserves with the same rigor as airworthiness (EASA, 2025).

Energy provisioning is a binding constraint across geographies. Fast-turnaround passenger operations and high-throughput cargo rotations drive charger count and power levels; recent NREL analyses for FAA indicate site peaks in the multimegawatt range (with scenario maxima on the order of ~10–13 MW per site) and recommend pairing grid upgrades with on-site storage and renewable generation to buffer load and improve resilience (NREL, 2023, 2024). Rural or exurban sites face different challenges—electrical distribution may be weaker or more distant—making modular storage and microgrids particularly valuable during early deployment (NREL, 2023, 2024).

- **Land use, zoning, and community fit:** Urban vertiports must be planned as civic infrastructure, not isolated pads. Contemporary planning guidance recommends integrating vertiports into comprehensive plans and zoning codes, employing tools such as overlay districts, separation distances from sensitive receptors (schools and hospitals), and time-of-day operating constraints to manage noise exposure and privacy concerns. Suburban siting can leverage park-and-ride lots, district centers, or hospital campuses; rural siting often co-locates with existing GA aerodromes or logistics hubs to minimize new land take while maximizing network reach. These planning instruments allow graduated access—permitting early services under well-defined conditions while data are gathered for subsequent rule refinement.
- **Airspace and operations: layering density with safety:** Operational concepts should scale from low-density corridors to networked, high-tempo nodes. In the near term, both FAA and EASA approaches support structured routes and procedures that minimize conflicts with existing traffic and critical infrastructure. FAA's *Innovate28* envisions initial integration at "key sites" using existing services and minimal changes, then progressively adding capabilities (e.g., higher throughput, expanded hours, additional weather minima) as safety cases and procedures mature—coherently with UAM ConOps 2.0 and the NAS integration portfolio (FAA, 2023b). In Europe, IAM sits alongside U-space deployments to provide digital services (strategic deconfliction, and dynamic reconfiguration) as operations densify—ensuring that procedures, ATC/UTM interfaces, and ground constraints (energy/turnaround) are co-optimized (EASA, 2025).
- **Demand and service design: matching use cases to geographies:** Demand patterns should drive topology. Urban cores tend to support airport-city shuttles and cross-town links where surface congestion is severe; suburbs benefit from feeder routes to rail/metro and cross-suburban links poorly served by legacy transit; rural regions can gain most from cargo logistics (medical and e-commerce) and essential-travel services that connect small towns to regional hubs. U.S. and state guidance frames AAM as an incremental addition to the mobility mix, with early services deliberately focused on a small number of reliable, repeatable missions before expanding the lattice of routes (FAA, 2023c; GDOT, 2024). For planning purposes, city-region analyses often assume short-haul mission radii (tens of miles) for frequent urban/suburban services, with longer intraregional legs in lower-density areas where air distances replace slow surface trips—an allocation that also balances battery endurance and reserve energy requirements (APA, 2024; FAA, 2023a).
- **Passenger and cargo: complementary scaling levers:** A mixed portfolio accelerates scale. Cargo operations (time-critical medical logistics, middle-/last-mile parcels) can validate charging strategies, ground flows, and UTM

links in lower-exposure settings, seeding infrastructure that later supports passenger air-taxi services. Conversely, passenger trunk routes at airports and major interchanges can anchor demand for high-capacity vertiports and justify grid upgrades that benefit cargo rotations during off-peak windows. Planning and siting guidance therefore encourages co-location and shared standards (markings, safety areas, and energy interfaces) so that passenger and cargo missions can time-share assets and evolve without stranded capital (FAA EB-105A; APA PAS).

- **Phasing strategy: from early access to robust networks:** Putting the pieces together, a robust scaling strategy typically proceeds as follows (Table 8.1):
  1. *Key-site activation* with limited routes (often airport-city) using retrofitted sites and conservative procedures;
  2. *Grid and charging uplift* at anchor vertiports, coupled with on-site storage to stabilize peaks;
  3. *Zoning codification* and performance-based operating conditions informed by measured noise/operations data;
  4. *Network extension* to suburban and then rural nodes, prioritizing cargo corridors with clear public benefit; and
  5. *Density upgrades* (longer hours, higher throughput, and expanded weather envelopes) enabled by U-space/UTM services and iterative safety cases. This approach is consistent with Innovate28's phased integration and with EASA's operations-first IAM guidance, giving cities, regions, and operators a common blueprint to scale safely and credibly.

8.3 REGULATORY FLEXIBILITY IN EMERGING MARKETS

Amin Vafadar and Tomislav Lang

Emerging markets are converging on a "learn-by-doing" approach for AAM: create tightly supervised regulatory sandboxes, use phased or incremental approvals rather than "all-at-once" rulemaking, and anchor decisions in risk-based, performance-oriented methods that can be transposed across borders. This pattern reflects ICAO's direction of travel—through the *Advanced Air Mobility Study Group* (AAM-SG) and recent ICAO papers promoting sandboxes as tools for evidence-based rulemaking—and provides pragmatic pathways where legacy civil-aviation frameworks are still maturing (ICAO, 2025a, 2025b).

- **Regulatory sandboxes as controlled testbeds:** Brazil offers one of the clearest state-level exemplars. In 2024–2025, the National Civil Aviation Agency (ANAC) launched a Regulatory Sandbox for Vertiports, selected consortia, and signed commitment agreements to build and test vertiport infrastructure and operating models under direct authority oversight for up to 24 months; in August 2025, ANAC codified the sandbox's legal basis by Resolution No. 775 (ANAC, 2023, 2025a, 2025b, 2025c). These instruments formalize the state's ability to relax or tailor rules temporarily, gather operational data (layout, approach/departure paths, fire & rescue, energy/charging, and access control), and then crystallize those findings into durable guidance

TABLE 8.1
Scaling eVTOL networks across geographies

Dimension	Urban	Suburban	Rural/Exurban	Planning Notes
Infrastructure & Energy	Retrofit heliports/rooftops; intermodal hubs at rail/airports; high charger density; grid peaks multi-MW; on-site storage to buffer peaks.	Park-&-ride, district centers, hospitals; moderate charger count; shared hubs with rail/metro feeders.	Co-locate at GA airfields/ logistics hubs; weaker grids ⇒ modular storage/ microgrids; fewer chargers.	Use FAA EB-105A geometry & markings; plan diversion sites; pair grid upgrades with storage.
Land Use & Zoning	Overlay districts, setbacks from sensitive receptors; time-of-day constraints; privacy/ noise protections.	Leverage existing centers/ parking; integrate with local plans.	Minimize new land take; leverage existing aerodromes/ logistics.	Phase access; gather ops/noise data to refine rules.
Airspace & Operations	Structured routes/corridors initially; scale to higher density with U-space/UTM; expanded hours/IFR as evidence accrues.	Feeder corridors to metro/ rail; progressive throughput upgrades.	Low-density corridors; weather/terrain considerations.	Crawl-walk-run integration aligned to Innovate28 & EASA IAM.
Demand & Use Cases	Airport-city shuttles; cross-town links where congestion is severe.	Rail/metro feeders; cross-suburban links.	Cargo logistics (medical/e-commerce), essential travel to regional hubs.	Short-haul frequent urban/ suburban vs. longer intraregional rural legs; balance battery reserves.
Passenger & Cargo	Co-located; passenger peaks justify high-capacity nodes; cargo uses off-peak windows.	Shared assets/time-sharing.	Cargo-first to seed infrastructure; passenger later.	Design for dual-use from day one to avoid stranded capital.
Phasing Strategy	Activate key sites; conservative procedures; iterate.	Extend to suburban nodes.	Add rural corridors; cargo-first.	Density upgrades via digital services; expand weather envelopes.

and amendments to existing regulations. Although industry press releases also describe winning teams and timelines, the primary sources are ANAC's own notices and presentations, underscoring this as a civil, regulator-led program rather than a vendor-led initiative (ANAC, 2025a).

The United Kingdom's CAA Regulatory Sandbox—used here as a transferable model—has similarly convened multiparty consortia (airports, ANSP, vertiport developers, and OEMs) to co-develop UAM Concepts of Operations for complex environments (e.g., London-Heathrow), with the CAA Innovation Hub explicitly aiming to identify regulatory gaps early and iterate (CAA, 2021, 2025). While the UK is not an "emerging" market, its sandbox mechanics—time-bounded trials; joint problem framing; regulator-observed demonstrations; and publication of lessons learned—map cleanly onto the needs of authorities building first-time AAM regimes.

- **Phased approvals and "crawl-walk-run" integration:** In lower-capacity regulatory environments, incremental certification allows authorities to admit narrow, low-risk operations first (e.g., piloted, VFR, and day) and widen privileges (night/IFR, higher density, and autonomy) as safety evidence accrues. Although articulated prominently by the FAA's *Innovate28* in a U.S. context, the concept is jurisdiction-agnostic and is increasingly echoed in national roadmaps and ANSP planning materials worldwide: start with essential services at designated key sites, instrument operations for data capture, and then codify extensions based on demonstrated performance (FAA, 2023b). Emerging-market authorities can adopt the same sequencing—often via exemptions, test-operation permits, or temporary operating specifications—while they build permanent rules.
- **Risk-based, performance-oriented methods:** To avoid "copy-and-paste" from legacy aircraft categories, many authorities are leaning on JARUS SORA v2.5, the de facto global methodology for Specific Operations Risk Assessment. SORA 2.5 updates ground- and air-risk models, introduces a cyber-safety extension, and clarifies documentation and Operational Safety Objectives—giving regulators a portable template to authorize novel AAM operations on a case-by-case basis while keeping safety intent intact (EASA/JARUS, 2023; JARUS, 2024). This is especially valuable for cargo eVTOL and RPAS logistics (medical supplies and middle-/last-mile), often used as first movers in emerging markets to de-risk passenger missions.
- **Public-private demonstrators under government orchestration:** South Korea's K-UAM Grand Challenge—a state-run, multiphase program—provides a replicable governance pattern: consortia flight-test aircraft, UTM/U-space services, and airport procedures under MOLIT supervision to validate integrated operations and inform certification pathways (MOLIT, 2023). Although Korea is an advanced economy, the *structure*—government-set objectives, staged trials, and publication of integration lessons—is directly applicable to regulators building capacity in newer markets.
- **Airspace readiness via corridor mapping and multiagency MoUs:** In the United Arab Emirates, the GCAA has publicly initiated the mapping of air

corridors for air taxis and cargo drones and is processing applications for commercial air transport certificates for eVTOL operators; Dubai's RTA, DCAA, and GCAA have coordinated exclusive concessions and site-specific deployments (corridors, vertiports, and operating windows) to enable early entry, then scale (Joby, 2025; Reuters, 2025a). The policy logic—pre-define corridors and nodes, then iterate requirements based on measured performance and local conditions (e.g., heat and dust)—is a generalizable technique for emerging markets with strong central planning capacity.

- **Foundational enablers and international scaffolding:** Beyond state experiments, international guidance helps keep national flexibility aligned with global interoperability. ICAO's AAM-SG is crafting a holistic framework and convening expert panels to channel AAM topics into SARPs, while development finance actors (e.g., World Bank toolkits) advise on performance-based regulation, PPP models, and institutional roles for safe civilian drone/AAM integration—resources that many emerging regulators directly adapt (ICAO, 2025a).
- **Synthesis:** Across Brazil, the UK model, Korea, and the UAE, a common architecture emerges: (i) sandbox real operations with explicit safety cases and data capture; (ii) phase approvals and privileges as evidence accumulates; (iii) use risk-based toolkits (SORA 2.5) to tailor mitigations; (iv) orchestrate public-private demonstrators under competent authority oversight; and (v) align with ICAO/EASA/UTM standards to preserve international portability. For civil AAM in emerging markets, this adaptive governance paradigm avoids waiting for perfect, comprehensive rules and instead earns permanent regulation through measured, transparent performance (Table 8.2).

8.4 SMART-CITY INTEGRATION AND INTERMODAL LINKS WITH AUTONOMOUS GROUND TRANSPORT

Amin Vafadar and Tomislav Lang

Integrating AAM into "smart cities" requires the coupling of digital platforms and physical hubs so that eVTOL services function as another layer of urban mobility rather than a standalone novelty. In Europe, this alignment is being structured via the SESAR 3/European ATM Master Plan and U-space demonstrations, which explicitly connect urban air operations, digital data exchange, and air-ground coordination; the 2025 Master Plan updates the U-space roadmap to reflect IAM needs and higher service levels (U3–U4), while very-large-scale demos like CORUS-XUAM validate procedures linking U-space services with conventional ATC in urban/suburban/airport environments (EUROCONTROL/SESAR, 2024; SESAR 3 JU, 2022; SESAR 3 JU, 2025).

- **Digital integration:** AAM must plug into the same information backbone that coordinates metros, buses, and on-demand road services. At Paris Pontoise-Cormeilles, Europe's first integrated vertiport terminal testbed (Groupe ADP/Skyports/RATP) was explicitly designed to test: (i) vehicle/

TABLE 8.2
Regulatory flexibility in emerging markets

Mechanism	Illustrative Jurisdictions/Programs	Key Instruments/Tools	What It Enables	Transferable Lessons
Regulatory sandboxes	Brazil ANAC Vertiport Sandbox (Res. 775/2025); UK CAA Sandbox.	Time-bounded trials; tailored rules; data capture plans.	Testing vertiport layouts, charging, procedures under authority oversight.	Regulator-led, evidence-based rules; publish lessons learned.
Phased approvals ("crawl-walk-run")	FAA Innovate28 model; echoed in national roadmaps.	Incremental privileges: piloted→IFR/night→higher density→autonomy.	Early low-risk ops while building safety evidence.	Gate expansions to measured performance.
Risk-based authorization	JARUS SORA v2.5; EASA alignment.	Ground/air risk models; cyber extension; OSOs.	Case-by-case approvals incl. cargo BVLOS.	Portable template that preserves safety intent.
Public-private demonstrators	K-UAM Grand Challenge (Korea).	Staged flight tests; state-set objectives.	Integrated validation of aircraft + UTM + procedures.	Government orchestration de-risks integration.
Airspace readiness & MoUs	UAE corridor mapping; Dubai concessions.	Predefined corridors/nodes; operating windows.	Faster initial service with local constraints (heat/dust).	Iterate based on measured performance.

charging integration on the apron; (ii) flight scheduling, situational awareness, and information exchange; and (iii) the end-to-end passenger journey with security, check-in, and SITA-enabled biometrics. Demonstration flights by Volocopter's 2X were run in concert with the terminal processes to exercise the digital chain from booking to boarding and flight monitoring (Future Travel Experience, 2022; Groupe ADP, 2022; SITA/Skyports, 2022; Skyports, 2022; Volocopter, 2022a). Volocopter's VoloIQ platform is positioned as the "digital backbone" for routing, fleet, and network optimization, and has been referenced as the underpinning of integrated operations at EU testbeds (Aviation Week, 2022; Volocopter, 2025).

This platform logic extends to urban Mobility-as-a-Service (MaaS). In Korea's state-run K-UAM program, Kakao Mobility (the country's dominant MaaS marketplace) is integrating eVTOL booking and multimodal trip planning, with public flight demonstrations linked to the Grand Challenge phases; its consortium (with LG Uplus and others) provides a concrete blueprint for MaaS-native AAM in which one app orchestrates first-/last-mile, vertiport processing, and the flight segment (AIN, 2024b; Archer Aviation & Kakao Mobility, 2024; Korea Herald, 2024). At the framework level, the MaaS Alliance sets interoperability expectations (marketplaces, APIs, and payments) that city authorities can reuse when adding an aerial mode to existing digital mobility ecosystems (MaaS Alliance, 2023).

- **Physical intermodality:** Smart-city integration is equally about co-location. The Paris testbed was conceived to trial passenger flows and charging operations in a terminal configuration that mirrors airport/rail hubs, anticipating deployment at stations, airports, and district centers (Groupe ADP, 2022; Skyports, 2022). The UK's Urban-Air Port "Air-One" (Coventry) prototyped a pop-up, off-grid vertiport powered by hydrogen fuel cells (AFC Energy) with on-site EV charging, a passenger lounge, and a cargo hub—hosting public demonstrations and ~150 cargo eVTOL missions during its run. The off-grid design illustrates how vertiports can operate as smart nodes that knit air mobility to ground EV ecosystems and can be sited temporarily to test demand or support events (UAP, 2024; AFC Energy case study; UAP launch news release; PwC UK review).
- **Autonomous ground transport links:** The intermodal vision pairs eVTOL flights with autonomous shuttles/robo-taxis to solve first-/last-mile access. Policy guidance from the World Economic Forum emphasizes integrating AVs with public transport backbones—precisely the model AAM requires to avoid duplicative road traffic—so that AVs feed passengers to vertiport-equipped hubs and distribute arrivals onward (WEF, 2024). Early operational patterns in Seoul (K-UAM) similarly coordinate airport vertiports (e.g., Gimpo) with city corridors along the Han River and MaaS platforms that can dispatch ground vehicles to vertiports, a template cities can adapt as AV fleets scale (AIN, 2024a; Archer Aviation & Kakao Mobility, 2024; Korea Herald, 2024).
- **Airspace-city systems coupling:** Digital twins and network-wide simulation are becoming core to planning. NASA and ATM communities use

digital-twin concepts to test proposed changes in traffic and procedures offline with real data—useful for anticipating eVTOL flows and emergency contingencies before scaling (NASA Technology, n.d.). In Europe, the CORUS-XUAM demos and SESAR work program document live trials where U-space services exchange data with ATC and operators, confirming that procedural and data interfaces can support urban operations and interfaces to major airports—an essential prerequisite for reliable, high-tempo city service (SESAR 3 JU, 2022, 2025).

- **Implications for cities and operators:** The practical pattern emerging from Paris, Coventry, and Seoul is consistent: (1) implement a unified digital stack (operator platform + vertiport OS + MaaS marketplace + U-space/ATC data exchange); (2) co-locate vertiports at intermodal nodes (airports, rail hubs, and district centers) with grid/charging and, where needed, off-grid storage or hydrogen power; (3) design AV-assisted first-/last-mile flows; and (4) instrument operations to feed noise/performance and throughput data back into U-space/ATM planning. When these elements are implemented, AAM behaves like a true smart-city subsystem: discoverable in the same trip planner, paid through the same wallet, using the same interchanges, and governed by the same data-sharing and performance regimes as surface transport (Table 8.3).

8.5 CASE STUDY: SOUTHERN EUROPE—eVTOL NETWORK EXPANSION

Amin Vafadar and Tomislav Lang

Southern Europe offers an instructive laboratory for scaling AAM because it combines airport-city trunk demand, tourism-driven inter-island flows, and public-interest logistics—all within the EASA regulatory orbit and the EU's U-space deployment roadmap. This case study examines three complementary strands: (i) Rome-Lazio (airport-city passenger trunking under an ENAC regulatory sandbox), (ii) Greece (tourism-led, multi-vertiport strategy across mainland and islands), and (iii) Spain (U-space-enabled medical logistics in dense urban airspace). Together, they illustrate how a coherent EU policy scaffold (EASA IAM; EU U-space) underpins growth paths that are adaptable across passenger and cargo use cases (EASA, 2025; EUROCONTROL, 2025).

8.5.1 Rome-Lazio (Italy): Airport-City Trunk Route under a Regulator-Supervised Sandbox

On October 6, 2022, Aeroporti di Roma (ADR), UrbanV, Volocopter, and partners inaugurated Italy's first fully functional AAM testing vertiport at Fiumicino (UV-0) and conducted the country's first crewed eVTOL public test flights—a five-minute sortie along a figure-eight profile at ~40 m AGL (AIN, 2024a; Volocopter, 2022a). The facility's electrical system was explicitly designed to trial multiple eVTOL charging modalities (fast charging and battery swap), and the site operates within a regulatory

TABLE 8.3
Smart-city integration and intermodal links

Integration Layer	Function in AAM	Programs/Testbeds	Interoperability Elements	Implementation Notes
Digital backbone (Operator + Vertiport OS)	Fleet/network optimization; charging/turnaround orchestration; safety monitoring.	VoloIQ at EU testbeds; Paris Pontoise terminal pilot.	APIs, booking, flight scheduling, status exchange	Treat as part of the city's MaaS data fabric.
MaaS/marketplaces	Trip planning, ticketing, payments, first-/last-mile orchestration.	Kakao Mobility in K-UAM; MaaS Alliance guidelines.	Open APIs; tariff & identity integration.	One-app experience across modes.
U-space/UTM↔ATC data exchange	Strategic deconfliction, conformance, dynamic reconfiguration.	CORUS-XUAM; U-ELCOME; SESAR roadmap.	Common info services; DSS/FIMS bridges.	Enable higher service levels (U3→U4) as density grows.
Physical intermodality	Co-location at airports/rail hubs/ district centers; off-grid resilience.	Paris testbed; Urban-Air Port "Air-One".	Wayfinding, security, passenger/cargo flows.	Design for modularity and pop-up pilots.
Autonomous ground links	Robo-taxis/shuttles feeding vertiports; curb/stand management.	WEF policy guidance; Seoul corridor demos.	Dispatch & curbside API coordination.	Align with public transport backbones to avoid duplication.

sandbox approved by ENAC, allowing progressive expansion of ground/flight test complexity (UrbanV, 2022). Strategically, the sandbox de-risks an airport-city passenger trunk (Fiumicino ↔ center), while accumulating the operational evidence needed for network extension to multiple Rome sites (UrbanV's stated plan: 10–12 vertiports in the metro) and, subsequently, to other Italian/French cities (UrbanV, 2022, 2023).

- **Implementation plan:** First, the airport anchor consolidates predictable demand (business/leisure air-rail substitution), justifying grid upgrades and high-throughput apron design that later benefit cargo rotations during off-peak hours. Second, sandbox governance allows ENAC to iterate procedures (routes, ground handling, and safety areas) while referencing EASA's IAM operations-first framework, compressing the rule-evidence loop (EASA, 2025). Third, by validating charging interoperability and terminal processes (e.g., integration with VoloIQ), the testbed creates portable templates for other Italian hubs (Volocopter, 2022b).

8.5.2 Greece: Tourism-Centric, Multi-Vertiport Network for Inter-Island Connectivity

In May–June 2024, Aria Hotels (a Libra Group subsidiary) announced a €50m program to construct and operate four vertiports across Greece, paired with leases for ten all-electric eVTOLs (1 pilot + 5 pax) positioned as "flying ferries" between the mainland and islands, with late-2026 service targets (AIN, 2024a; Libra Group, 2024; Travel Weekly, 2024). The plan emphasizes charging infrastructure and passenger amenities with a small land footprint, explicitly oriented to tourism flows and premium time savings over ferries. Operationally, such a lattice can carry passengers by day and high-value cargo (e.g., hospitality logistics and medical consignments) by night, exploiting the same assets under differing duty cycles. While Greece's AAM framework will ultimately align to EASA IAM and EU U-space constructs, the private-sector-led, regulator-coordinated model offers a pragmatic route to early services in geographically fragmented markets (AIN, 2024b; Libra Group, 2024).

- **Implementation plan:** Tourism corridors with recurrent seasonal demand can underwrite the capex for a cluster of vertiports, after which adjacent public-service use cases (e.g., inter-island medical logistics) can be layered in without new groundworks. The network also creates data feedback (load factors, weather, and charging cycles) to refine siting and operating windows for subsequent islands or mainland spokes (AIN, 2024b; Libra Group, 2024).

8.5.3 Spain: U-Space Deployment Enabling Medical Logistics in Complex Urban Airspace

Spain provides a cargo-first complement via U-space. In September 2025, ENAIRE (Spain's ANSP and designated Common Information Services Provider) and partners conducted medical-supply drone deliveries between Hospital Carlos III and Hospital La Paz in Madrid, within a U-space trial under the EU-funded U-ELCOME project coordinated by EUROCONTROL. The three-minute urban flight operated near Barajas and existing heliports, using U-space services for flight planning, real-time tracking, and conformance monitoring; AESA operational authorization, SORA-based analysis, and layered safety measures (parachute, FTS, and containment) were cited (ENAIRE, 2025a; ENAIRE PDF, 2025b; EUROCONTROL U-ELCOME, 2025). Related U-ELCOME validations in the Canary Islands exercised emergency reconnaissance and medicine transport in controlled airspace at the Fuerteventura Stratoport (ENAIRE, 2025c).

- **Implementation plan:** U-space-enabled logistics (medical, laboratory, and time-critical spares) can pre-figure passenger eVTOL operations by establishing digital services, procedures, and public-sector value cases in dense airspace—lowering barriers for later mixed (passenger + cargo) networks (ENAIRE, 2025a; EUROCONTROL, 2025).

8.5.4 Cross-Cutting Lessons for Southern Europe

Following up on the theory in Chapter 8 and the presented case studies, we conclude the following lessons for the implementation of AAM in Southern Europe (Table 8.4):

1. Anchor at airports; branch into city and tourism nodes. Rome shows how airport-city trunks can finance grid-intensive hubs; Greek inter-island networks turn seasonal peaks into durable infrastructure footprints (Libra Group, 2024).
2. Use sandboxes and U-space as growth engines. ENAC's sandbox accelerates passenger trunking; U-space demonstrators (Madrid and Fuerteventura) scale digital/operational readiness for cargo that later generalizes to passenger operations (ENAIRE, 2025a, 2025c).
3. Design for dual use from day one. Charging, apron, and terminal flows should assume day-night role switching (passenger ↔ cargo) to protect capital productivity and smooth energy peaks (UrbanV, 2022).
4. Keep alignment with EASA IAM and EU U-space. A common operations-first framework (IAM) and staged deployment of U-space services make local pilots portable across borders—critical for multi-state operators in Southern Europe (EASA, 2025; EUROCONTROL, 2025).

8.6 CONCLUSION

Dimitrios Ziakkas, Amin Vafadar, and Debra Henneberry

AAM networks scale when infrastructure, policy, operations, and demand are co-designed and phased against evidence. The practical blueprint that emerges across geographies and governance models is consistent: start small at high-value nodes; codify community-fit through zoning; invest early in grid/charging with storage; operate on structured routes tied into U-space/UTM; mix passenger and cargo to lift utilization; and expand hours, weather minima, and throughput as safety cases mature. Regulatory sandboxes, SORA-based authorizations, and government-orchestrated

TABLE 8.4
Southern Europe—network expansion patterns

Sub-Region/ Program	Primary Focus	Anchor Strategy	Governance Tooling	Why It Scales	Notes
Rome-Lazio (Italy)	Airport-city passenger trunk under sandbox.	FCO UV-0 vertiport; charging (fast + swap); VoloIQ-linked terminal processes.	ENAC sandbox; EASA IAM alignment.	Airport demand justifies grid/ apron capacity; portable procedures to other cities.	Target metro network (10–12 sites).
Greece (tourism corridors)	Inter-island passenger network; dual-use with cargo.	Multi-vertiport strategy at tourism nodes; small land footprint.	Private-led, regulator-coordinated; EU U-space backdrop.	Seasonal peaks underwrite capex; night cargo uses off-peak windows.	Data feedback on load/ weather/charging for expansion.
Spain (Madrid & Canary Islands)	U-space logistics (medical supplies/emergency reconnaissance).	Hospital-to-hospital urban corridors; stratoport trials.	ENAIRE CIS; U-ELCOME; SORA-based authorizations.	Builds digital services & procedures that later generalize to passenger ops.	Dense airspace validation near major airport.

demonstrations shorten the distance between promising trials and permanent rules, especially in emerging markets. Smart-city integration—digital plus physical—turns eVTOL from a novelty into a dependable layer in the mobility stack.

Finally, scaling choices should be read through a business and benefits lens: price architecture, utilization, CASK/CASM versus RASK, payback, and scenario-based ROI guide where (and how fast) to build; societal metrics—jobs, access, reliability, emissions, and resilience—sustain public and political support. Cities and operators that instrument operations and publish performance will earn the privileges that allow density to rise—safely, credibly, and with visible value delivered to communities.

REFERENCES

AIN Online. (2024a, May 29). *Aria Hotels to establish four vertiports in Greece (leases 10 eVTOLs)*. https://www.ainonline.com/aviation-news/business-aviation/2024-05-29/aria-hotels-establish-four-vertiports-greece

AIN Online. (2024b, May 30). *Archer joins K-UAM Challenge with Kakao Mobility*. https://www.ainonline.com/aviation-news/futureflight/2024-05-30/archer-joins-korean-uam-challenge-kakao-mobility

ANAC (Agência Nacional de Aviação Civil—Brazil). (2023). *Sandbox Vertiportos/EVTOLs (programme overview)*. https://www.gov.br/anac/pt-br/assuntos/sandbox-regulatorio/sandbox-vertiportos-evtols

ANAC (Agência Nacional de Aviação Civil—Brazil). (2025a). *Edital nº 3/SIA/2025*. https://www.anac.gov.br/assuntos/legislacao/editais/2025/edital-no-3-sia-2025

ANAC (Agência Nacional de Aviação Civil—Brazil). (2025b, April 11). *ANAC firma termos de admissão para desenvolver projetos e regulação para vertiportos*. https://www.gov.br/anac/pt-br/noticias/2025/anac-firma-termos-de-admissao-para-desenvolver-projetos-e-regulacao-para-vertiportos

ANAC (Agência Nacional de Aviação Civil—Brazil). (2025c, August 5). *Resolução nº 775—Sandbox rules*. https://www.anac.gov.br/assuntos/legislacao/legislacao-1/resolucoes/2025/resolucao-775

Archer Aviation, & Kakao Mobility. (2024, May 30). *Kakao Mobility selects Archer Aviation as its eVTOL partner and will fund Archer's Korean commercialization efforts*. https://investors.archer.com/news/news-details/2024/KakaoMobility-Selects-Archer-Aviation-As-Its-eVTOL-Partner-And-Will-Fund-Archers-Korean-Commercialization-Efforts-Together-They-Will-Conduct-Public-Flight-Demonstrations-As-Part-Of-Koreas-Grand-Challenge-As-Soon-As-Q42024/default.aspx

Aviation Week. (2022, October 6). *Italy inaugurates first vertiport; VoloIQ cited as digital backbone*. https://aviationweek.com/aerospace/advanced-air-mobility/italy-inaugurates-first-vertiport-evtol-test-flight

CAA (UK Civil Aviation Authority). (2021). *Future Air Mobility Regulatory Sandbox—Eve case study*. https://www.caa.co.uk/publication/download/19168

CAA (UK Civil Aviation Authority). (2025). *Enabling Advanced Air Mobility—Regulator innovation pages*. https://www.caa.co.uk/our-work/research-and-innovation/enabling-advanced-air-mobility/

EASA (European Union Aviation Safety Agency). (2025, July 9). *EASA steps up regulatory framework for Innovative Air Mobility (IAM)*. https://www.easa.europa.eu/en/newsroom-and-events/news/easa-steps-regulatory-framework-innovative-air-mobility

EASA/JARUS. (2023). *SORA 2.5—Summary of changes* [Slide deck]. https://www.easa.europa.eu/sites/default/files/dfu/sora_workshop_feb_2023_-_4_summary_of_changes_sora.pdf

ENAIRE. (2025a, September 17). *Medical supply drone completes demonstration flight in Madrid*. https://www.enaire.es/en_GB/2025_09_17/ndp_gb_successful_drone_flight_carrying_medical_supplies_madrid

ENAIRE. (2025b, September 23). *Successful drone flight carrying medical supplies in Madrid* [Press PDF]. https://www.enaire.es/docs/download/es_ES/press_release/PDF_GB_Successful_drone_flight_carrying_medical_supplies_Madrid.pdf

ENAIRE. (2025c, September 26). *European U-ELCOME final validation drone flight demo in Fuerteventura*. https://www.enaire.es/en_GB/2025_09_26/ndp_gb_european_uelcome_final_drone_flight_demo_fuerteventura

EUROCONTROL. (2024, December 12). *2025 edition of the European ATM Master Plan sets clear path towards digital and green European sky*. https://www.eurocontrol.int/news/2025-edition-european-atm-master-plan-sets-clear-path-towards-digital-and-green-european-sky

EUROCONTROL. (2025). *U-ELCOME project overview*. https://www.eurocontrol.int/project/u-elcome

FAA (Federal Aviation Administration). (2023a). *Advanced Air Mobility (AAM) implementation plan—Innovate28 (v1.0)*. https://www.faa.gov/sites/faa.gov/files/AAM-I28-Implementation-Plan.pdf

FAA (Federal Aviation Administration). (2023b). *AAM implementation plan (summary page)*. https://www.faa.gov/air-taxis/implementation-plan

FAA (Federal Aviation Administration). (2023c, October 4). *Integrating AAM in the NAS* [REDAC briefing]. https://www.faa.gov/about/office_org/headquarters_offices/ang/redac/redac-full-202310-IntegratingAamNas.pdf

FAA (Federal Aviation Administration). (2024, December 27). *Engineering brief 105A: Vertiport design (supplement to AC 150/5390-2D)*. https://www.faa.gov/airports/engineering/engineering_briefs/eb_105a_vertiports

Future Travel Experience. (2022). *Groupe ADP & Skyports launch vertiport passenger terminal testbed in Paris*. https://www.futuretravelexperience.com/2022/11/groupe-adp-and-skyports-launch-vertiport-passenger-terminal-testbed-in-paris/

GDOT (Georgia Department of Transportation). (2024). *AAM community guidebook*. https://www.dot.ga.gov/InvestSmart/Aviation/AAM/Community%20Guidebook%20-%20GDOT%20AAM%20Study.pdf

Groupe ADP. (2022, November 10). *First integrated vertiport passenger terminal testbed in Europe (Pontoise)*. https://presse.groupeadp.fr/first-vertiport-pontoise/

ICAO (International Civil Aviation Organization). (2025a). *Advanced Air Mobility Study Group (AAM-SG)—Scope & activities*. https://www.icao.int/UA/advanced-air-mobility-study-group-aam-sg

ICAO (International Civil Aviation Organization). (2025b). *Promoting regulatory sandboxes as tools for evidence-based regulation* [Working paper]. https://www.icao.int/sites/default/files/Meetings/a42/Documents/WP/wp_217_en.pdf

JARUS (Joint Authorities for Rulemaking on Unmanned Systems). (2024). *SORA v2.5—Main body & annexes*. https://jarus-rpas.org/publications/

Joby Aviation. (2025, June 30). *Dubai programme and six-year exclusivity (RTA/DCAA/GCAA)*. https://www.jobyaviation.com/news/joby-cements-global-lead-in-air-taxi-industry/

Korea Herald. (2024, November 11). *Seoul to begin UAM demonstration services in 2025*. https://www.koreaherald.com/article/3851964

Libra Group. (2024, May 28). *Aria Hotels to construct eVTOL vertiports (Greece)*. https://www.libra.com/news-and-insights/aria-hotels-to-construct-evtol-vertiports/

MaaS Alliance. (2023, October 2). *Mobility marketplaces & MaaS (White paper)*. https://maas-alliance.eu/wp-content/uploads/2023/10/White-paper_Mobility-Marketplaces-and-MaaS.pdf

MOLIT (Ministry of Land, Infrastructure and Transport, Korea). (2023). *K-UAM grand challenge—Programme overview* [PDF]. https://www.molit.go.kr/LCMS/DWN.jsp?fileName=K-UAM+Grand+Challenge.pdf&fold=%2Feng0301%2F

NASA Technology. (n.d.). *Digital twin NAS simulator*. https://technology.nasa.gov/tags/simulation

NREL (National Renewable Energy Laboratory). (2023). *Federal Aviation Administration vertiport electrical infrastructure study (NREL/TP-5R00-86245)*. https://www.nrel.gov/docs/fy24osti/86245.pdf

NREL (National Renewable Energy Laboratory). (2024, March 7). *Electric aircraft will need powerful ports*. https://www.nrel.gov/news/detail/program/2024/electric-aircrafts-will-need-powerful-ports

Reuters. (2025a, February 13). *UAE begins mapping corridors for air taxis and cargo drones*. https://www.reuters.com/world/middle-east/uae-begins-mapping-corridors-air-taxis-cargo-drones-2025-02-13/

Reuters. (2025b, July 1). *Dubai aims to beat the traffic with 2026 Joby air taxi liftoff*. https://www.reuters.com/world/middle-east/dubai-aims-beat-traffic-with-2026-joby-air-taxi-liftoff-2025-07-01/

SESAR 3 Joint Undertaking. (2022). *Demonstrating everyday benefits of U-space*. https://www.sesarju.eu/sites/default/files/documents/reports/Uspace_May2022__FINAL.pdf

SESAR 3 Joint Undertaking. (2025). *European ATM Master Plan 2025 (U-space & IAM roadmap)*. https://www.sesarju.eu/sites/default/files/documents/reports/SESAR%20Master%20Plan%202025.pdf

SITA, & Skyports. (2022). *Skyports partners with SITA to explore technology for the AAM passenger journey*. https://skyports.net/skyports-partners-with-sita-to-explore-technology-for-the-aam-passenger-journey/

Skyports. (2022, November 10). *Vertiport testbed inaugurated in Paris (scope of testing)*. https://skyports.net/vertiport-testbed-for-european-urban-air-mobility-testing-inaugurated-in-paris/

Travel Weekly. (2024, June 7). *Electric aircraft to ferry tourists to Greek islands*. https://travelweekly.co.uk/news/electric-aircraft-to-ferry-tourists-on-short-hops-to-greek-islands

Urban Air Port. (2024, April 24). *From prototype to commercialisation—Air-One results (incl. ~150 cargo eVTOL flights, EV charging)*. https://www.urbanairport.com/uap-blog/from-prototype-to-commercialisation-airone-2-years-on

UrbanV. (2022). *UV-0 test vertiport (EN/IT pages; sandbox context)*. https://www.urbanv.com/en/uv-0-test-vertiport/

UrbanV. (2023). *UrbanV CEO outlines vision for Rome network (10–12 vertiports)*. https://www.urbanv.com/en/urbanv-ceo-outlines-vision-for-rome-network/

Volocopter. (2022a). *Vertiport testbed for EU UAM—End-to-end passenger journey & data exchange*. https://www.volocopter.com/en/newsroom/vertiport-testbed-for-eu-uam-paris

Volocopter. (2022b, October 6). *Italy's first vertiport deployed at Fiumicino Airport (flight profile; ENAC/ENAV clearances)*. https://www.volocopter.com/en/newsroom/italys-first-vertiport-deployed-at-fiumicino-airport

Volocopter. (2025). *VoloIQ—Digital backbone for UAM operations*. https://www.volocopter.com/en/solutions/voloiq

World Economic Forum. (2024). *How will autonomous vehicles shape future urban mobility?* https://www.weforum.org/stories/2024/10/how-will-autonomous-vehicles-shape-urban-mobility/

9 Environmental Sustainability, Cybersecurity, and Ethical Considerations Perspectives

Dimitrios Ziakkas and Amin Vafadar

9.1 INTRODUCTION

The transformation of Advanced Air Mobility (AAM) into a viable transport ecosystem represents more than a technological milestone; it is a test of how innovation can remain accountable to human values. As eVTOL aircraft and autonomous flight systems mature, sustainability, cybersecurity, and ethics have emerged as the *interlocking conditions* for legitimacy. The European Union Aviation Safety Agency (EASA) (EASA, 2022) defines this interdependence through its *Trustworthiness Framework*, which positions ethics as an operational layer within aviation governance rather than a philosophical add-on. The framework identifies four foundational principles—respect for human autonomy, prevention of harm, fairness, and explicability—that must underpin every phase of Artificial Intelligence (AI) deployment, from system design to post-market surveillance. Within AAM, these principles converge to form a new safety culture where *transparency and accountability* are as critical as aerodynamic stability or propulsion reliability.

Historically, aviation regulation has evolved around technical safety margins; however, the rise of autonomous and data-driven flight expands the moral perimeter of responsibility. Each eVTOL platform functions as a cyber-physical organism, interacting continuously with vertiports, digital traffic-management systems, and cloud-based maintenance networks. The resulting web of algorithmic decision-making demands assurance that machines will act in ways consistent with human intentions and societal norms (EASA, 2024b). The ethical imperative of "meaningful human control"—central to the *EASA (2022)* framework—becomes a prerequisite for trust in automation. This requirement extends to both *human-in-the-loop* and *human-on-the-loop* architectures, ensuring that even in highly autonomous flight, ultimate accountability remains human.

DOI: 10.1201/9781003685555-9

At the same time, AAM's promise of sustainability challenges the sector to align technological efficiency with planetary stewardship. Electrified propulsion, renewable-energy integration, and circular battery economies hold the potential to reduce life cycle greenhouse gas (GHG) emissions by up to 40% relative to comparable rotorcraft (ICCT, 2024). Yet environmental performance alone cannot secure public confidence; communities increasingly evaluate innovation through moral as well as technical lenses. The EASA (2022) report emphasizes that *trust in AI-enabled aviation is relational*—it grows when citizens perceive that technologies respect their autonomy, protect them from harm, distribute benefits fairly, and can be explained in understandable terms. These ethical expectations therefore shape market acceptance as decisively as cost or convenience (NASEM, 2024; OECD, 2024).

Cybersecurity further entwines ethics and safety. As aircraft systems migrate to cloud architectures and 5G-enabled data links, digital vulnerabilities become potential safety hazards. The *EASA AI Roadmap 2.0* (2024b) expands the ethical framework into practical assurance levels for machine-learning applications, demanding traceability of data sets, robustness against adversarial interference, and continuous oversight through algorithmic audits. In this sense, cyber-resilience becomes a moral obligation: protecting human life now includes protecting the integrity of the data and algorithms that sustain it.

Global governance bodies echo this integrative approach. The *International Civil Aviation Organization (ICAO, 2023a)* and *Federal Aviation Administration (FAA, 2023a)* increasingly frame sustainability and digital trust as mutually reinforcing. ICAO's *Circular 344* and Annex 16 link environmental protection to community well-being, while the FAA's *Innovate28* strategy mandates cybersecurity assessment alongside environmental reviews for vertiport projects. Across these initiatives runs a common thread first articulated by *EASA (2022)*—that public trust depends on a visible chain of accountability connecting ethical intention, technical design, and societal outcome.

This chapter builds upon that insight. Section 9.2 examines how carbon management, noise abatement, and circular-economy principles operationalize environmental responsibility. Section 9.3 addresses the expanding frontier of cybersecurity and AI safety, interpreting digital resilience through the lens of human autonomy and explicability. Section 9.4 develops the ethical and societal dimensions of AAM, integrating the EASA Trustworthiness Framework with global public-engagement models. Section 9.5 presents a comparative case study of regulatory innovations in Europe, the United States, Japan, and the UAE, culminating in Figure 9.1—a conceptual diagram of sustainable integration. The concluding section synthesizes these domains into a human-centric vision of responsible flight.

By grounding sustainability and security in ethics, AAM can evolve from a disruptive technology into a moral enterprise of mobility. The lesson from *EASA (2022)* is unambiguous: trustworthy aviation is not achieved when machines become flawless, but when their creators remain answerable to human purpose. The following sections explore how this principle can guide AAM toward a future where innovation, responsibility, and trust ascend together.

9.2 ENVIRONMENTAL IMPACT: CARBON FOOTPRINT, NOISE ABATEMENT

Dimitrios Ziakkas, Anastasia Kosmidou, and Andreas Diamantopoulos

9.2.1 Toward Carbon-Neutral Advanced Air Mobility

The environmental sustainability of AAM systems is inseparable from their public legitimacy. Although electrification promises to decouple mobility growth from carbon emissions, the actual carbon footprint of eVTOL operations depends on the full life cycle of materials, manufacturing, energy sourcing, and end-of-life management. The aviation industry's contribution to anthropogenic climate forcing—estimated at 3.5% of total radiative forcing (Lee et al., 2021)—has driven regulators to explore pathways to net-zero emissions by 2050. Within this framework, AAM offers a laboratory for sustainable aviation: its small airframes, short mission profiles, and integration with renewable grids make decarbonization technically achievable within a single-design generation (EASA, 2024a; ICAO, 2023a).

Recent life cycle assessments (LCA) suggest that, when powered by electricity from grids exceeding 50% renewable content, eVTOL aircraft can reduce total GHG emissions by 30–40% compared to turbine helicopters of similar payload (ICCT, 2024). NASA's AAM Integration Study (2024) reports average operational energy intensities of 120–150 Wh per passenger-kilometer, less than one-third of those for conventional rotorcraft. Yet these improvements are conditional on upstream factors: battery chemistry, energy storage degradation, and recycling logistics can offset gains if not properly managed (Zhao et al., 2023). Consequently, environmental planning for AAM must extend beyond propulsion to embrace circular-economy principles—closed-loop material recovery, responsible mining of lithium and cobalt, and second-life applications for degraded cells (IATA, 2023).

9.2.2 Energy Systems and Supply Chain Sustainability

The scalability of AAM will be constrained by the carbon intensity of local power grids. The National Renewable Energy Laboratory (NREL, 2023) demonstrated that eVTOL networks operating from renewable-integrated vertiports could achieve 75–90% reductions in CO_2e compared with jet-A-fueled helicopters. However, if electricity is predominantly fossil-based—as in several rapidly growing economies—the environmental break-even point may not occur until flight volumes justify investment in cleaner grids. Therefore, governments view energy policy alignment as a precondition for sustainable AAM deployment (OECD, 2024).

Supply chain transparency is another determinant. The European Battery Regulation mandates carbon-footprint declarations for all large traction batteries from 2025, establishing benchmarks that eVTOL manufacturers must satisfy for market entry in the EU. Similarly, the U.S. Inflation Reduction Act (2022) links tax incentives for advanced aircraft to domestic content and low-carbon manufacturing. Such instruments reveal a geopolitical dimension: sustainability becomes not only an environmental necessity but also a competitive industrial policy influencing where eVTOL value chains are anchored (DOE, 2024).

9.2.3 Noise Abatement and Community Well-Being

Noise exposure remains the most tangible environmental concern for communities. Traditional helicopters generate effective perceived noise levels (EPNdB) between 90 and 95, whereas prototype eVTOLs record 65–70 EPNdB during takeoff and landing—roughly equivalent to urban background traffic (EASA, 2023). This 60–70% reduction results from multiple small rotors operating at higher frequencies and lower tip speeds. Yet psychoacoustic perception complicates the picture: tonal components above 1 kHz may produce irritation even at lower decibel levels (Stansfeld & Clark, 2015).

The International Civil Aviation Organization's Annex 16 Volume I establishes standardized noise-certification procedures, while EASA CS-VTOL.500 introduces mission-based metrics that evaluate not only maximum sound exposure but also community noise footprint— the spatial area experiencing >60 dB (A) events. Paris's *Reinventing Air Mobility* trials show that routing optimization and dynamic vertiport scheduling can reduce exposure zones by 25% without technological modification (ADP Group, 2024). Moreover, active-noise-control systems integrated into rotor shrouds have achieved an additional 5–6 dB reduction in prototype testing (Airbus, 2023).

9.2.4 Comparative Assessment of Global Approaches

Different regulatory ecosystems adopt varying strategies to quantify and mitigate AAM's environmental impact (Table 9.1). The EASA integrates environmental certification directly into its type-validation process, ensuring that sustainability metrics are treated alongside safety. The Federal Aviation Administration (FAA), through

TABLE 9.1
Comparative environmental indicators of AAM deployment (2024–2040)

Region/ Agency	Primary Focus	Carbon-Reduction Mechanism	Noise-Mitigation Strategy	Key Regulatory Instrument
EASA (EU)	Life cycle emissions, battery recycling	Renewable-based charging (≥70% grid mix), closed-loop materials	Rotor-frequency optimization and mission-based noise metrics	CS-VTOL Subpart E & EASA Environmental Certification Specifications 2024
FAA (USA)	Vertiport electrification, environmental impact assessments	Low-carbon grid integration and tax credits for green energy	Land use planning to avoid sensitive zones	FAA Order 1050.1 F Environmental Policy & Innovate28 Plan
METI (Japan)	Hydrogen hybrid propulsion, composites recycling	Hydrogen production from renewables and battery reuse	Urban noise maps and public consultation	AAM Roadmap 2040 (METI, 2024)
GCAA (UAE)	Integration with national ESG strategy	Solar-powered vertiports and offset programs	Flight-path design to minimize residential overflight	UAE Green Logistics Strategy 2030 (GACA, 2024)

its *Innovate28* initiative, addresses sustainability primarily via infrastructure electrification and environmental assessments at the vertiport level. In the Asia-Pacific region, Japan's Ministry of Economy, Trade and Industry (METI) emphasizes the development of hydrogen-hybrid propulsion and recyclable composite materials as long-term decarbonization pathways (METI, 2024). The United Arab Emirates (UAE) integrates noise and emissions standards within its broader *Green Logistics Strategy 2030*, linking AAM with national sustainability targets (GACA, 2024).

9.3 CYBERSECURITY CHALLENGES AND AI SAFETY IN AAM OPERATIONS

Dimitrios Ziakkas, Eva Maleviti, and Konstantinos Pechlivanis

9.3.1 The Digital Dependence of Advanced Air Mobility

As eVTOL aircraft evolve from isolated vehicles into networked nodes of urban airspace, their dependence on digital connectivity introduces a new class of risks. Each aircraft in an AAM ecosystem communicates continuously with vertiports, cloud-based traffic-management servers, energy-grid controllers, and maintenance databases. Such interconnectedness yields operational efficiency but expands the attack surface of aviation systems traditionally protected by physical isolation. A cyber-intrusion into one node could compromise not only a single aircraft but also the situational integrity of an entire urban air corridor (FAA, 2023b; NASEM, 2024).

The traditional safety paradigm of "fail-safe" engineering must therefore coexist with a "secure-by-design" philosophy. Research by the *European Union Agency for Cybersecurity (ENISA, 2023)* underscores that AAM systems merge three critical infrastructures—aviation, energy, and telecommunications—and any breach at their intersection could cascade across sectors. For instance, ransomware directed at vertiport charging networks could disrupt hundreds of scheduled flights within minutes, while manipulated navigation data could precipitate mid-air conflicts. The World Economic Forum (2025) classifies such digital interdependence as one of the principal vulnerabilities in the transition to sustainable mobility.

9.3.2 Regulatory Responses and Institutional Frameworks

Recognizing these risks, regulators have begun integrating cyber-resilience into the certification and oversight of AAM operations (Table 9.2). The FAA requires that new aircraft architectures include system-level security assessments consistent with *Advisory Circular AC 119–1 on Aircraft Systems Information Security Protection* (FAA, 2023a). In Europe, the EASA has established a cybersecurity rule set under *Commission Regulation (EU) 2023/203* that applies to both traditional aircraft and emerging eVTOL systems. These provisions oblige manufacturers to demonstrate threat modeling, encryption standards, and update-management capabilities before type certification (EASA, 2024a).

At the international level, ICAO Annex 17 and the *Aviation Cybersecurity Strategy 2020–2030* articulate a global baseline for cooperation, encouraging States to establish

TABLE 9.2
Global regulatory approaches to cybersecurity and AI safety in AAM (2024)

Region/ Regulator	Key Focus Area	Security Mechanisms Required	AI-Safety Requirements	Principal Reference
EASA (EU)	Integrated cyber-safety and AI governance	Threat modeling, encrypted data links, patch-management life cycle	Explainable AI (XAI) demonstrations; human-oversight plan	*EASA Artificial Intelligence Roadmap 2.0 (2024)*
FAA (USA)	System information security protection	Compliance with AC 119–1 and NIST SP 800–53 controls	AI risk assessment and continuous learning monitoring	*FAA Innovate28 Plan (2023a)*
MLIT (Japan)	Secure AAM network operations	End-to-end encryption; redundant command links	Human-on-the-loop requirement for autonomous flights	*MLIT AAM Cybersecurity Guideline (2024)*
GCAA (UAE)	Cyber risk integration with airspace management	Centralized incident reporting and AI-driven anomaly detection	Certification of AI functions through regulatory sandbox tests	*UAE AAM Safety and Security Framework (2024)*

national aviation-cybersecurity programs aligned with the *Montreal Declaration (2019)* (ICAO, 2023a). Meanwhile, Japan's Ministry of Land, Infrastructure, Transport and Tourism (MLIT) has issued a *Guideline for Secure Operation of Unmanned Aircraft and AAM Systems (2024)* requiring encrypted command links and multifactor authentication for all remotely piloted or automated flights. Together, these frameworks signal a shift from voluntary best practice to mandatory cyber certification as a prerequisite for market entry.

9.3.3 Artificial Intelligence Safety and Explainability

Cybersecurity and AI safety are converging disciplines within AAM. AI systems govern trajectory optimization, detect-and-avoid functions, maintenance prediction, and even passenger-interaction interfaces. While these systems can enhance safety by mitigating human error, they also introduce opaque decision-making processes that complicate certification and accountability (Bekele et al., 2023). The EASA AI Roadmap 2.0 defines four assurance levels for machine-learning applications, requiring traceability of data sets, interpretability of algorithms, and demonstrable robustness against adversarial attacks (EASA, 2024b).

To meet these criteria, manufacturers are adopting Explainable AI (XAI) models that allow regulators to audit the logic of automated decisions. For example, Airbus's *Project Vertex* employs hybrid neural-symbolic systems whose decision trees can be inspected during certification (Airbus, 2023). NASA's *Autonomy Integrity Verification*

Program (2024) similarly uses digital twins to simulate how AI flight controllers respond under degraded sensor conditions, thereby validating algorithmic reliability before deployment. These methods operationalize the ethical requirement of human accountability: operators must be able to explain the outcome of an AI-assisted flight event to authorities and affected individuals (Floridi et al., 2021).

9.3.4 Resilience and Incident Response

Beyond prevention, AAM must institutionalize cyber-resilience—the ability to detect, isolate, and recover from digital compromise while maintaining essential safety functions. The *ENISA Aviation Threat Landscape Report (2023)* notes that average incident-detection times in aviation systems exceed 150 days, underscoring the need for automated intrusion detection and blockchain-based log integrity. In the United States, the *Cybersecurity and Infrastructure Security Agency (CISA)* collaborates with NASA on the *AAM Cyber Resilience Program 2025*, which aims to shorten detection-to-response intervals to under 24 hours.

A human-centric perspective also recognizes that the weakest link in digital security is often behavioral. Continuous training for pilots, maintenance crews, and vertiport operators is essential to maintain a "security culture" analogous to aviation's safety culture. EASA (2024a) recommends that cyber-awareness modules become mandatory in AAM pilot and dispatcher certification syllabi. Embedding such awareness ensures that cybersecurity is experienced as a shared professional ethic rather than an external compliance burden.

9.3.5 The Human-Centric Integration of Digital Safety

Cybersecurity and AI safety are not purely technical domains but expressions of human responsibility in a digitized environment. AAM's legitimacy will depend on public confidence that automation enhances rather than erodes human control. Surveys by *NASEM (2024)* reveal that transparency about data protection and algorithmic oversight significantly increases willingness to use eVTOL services. When users understand how their information is secured and how AI decisions are supervised, perceived risk declines even more than with purely technical improvements.

Therefore, future regulation must evolve from prescriptive checklists to adaptive trust frameworks—systems of continuous learning where safety assurance grows with operational data. In this evolving paradigm, human factors specialists, ethicists, and cybersecurity engineers share a single mandate: to preserve trust, autonomy, and accountability in the digital cockpit.

9.4 ETHICAL CONSIDERATIONS AND COMMUNITY ACCEPTANCE

Dimitrios Ziakkas, Ioanna Lekea, and Konstantinos Pechlivanis

9.4.1 From Technical Compliance to Ethical Legitimacy

The emergence of AI-enabled eVTOL systems compels aviation to confront questions that transcend safety certification. While traditional regulation has focused on mechanical reliability and pilot competency, autonomy introduces dilemmas of accountability, transparency, and social justice. The EASA (2024a) emphasizes that ethical legitimacy—rather than mere legal conformity—constitutes the decisive factor for sustainable trust in AI. In practice, ethics becomes an *operational requirement*, shaping how AI is designed, validated, and explained to the public.

Within AAM, ethical assurance intersects directly with public acceptance. Autonomous flight control, passenger-data analytics, and predictive maintenance all rely on data ecosystems that must respect privacy and human agency. Surveys across pilot cities such as Paris, Los Angeles, and Osaka show that citizens' willingness to adopt AAM services depends as much on *perceived fairness and transparency* as on cost or safety (NASEM, 2024; OECD, 2024). Hence, ethical design is inseparable from market success: trust functions as an economic asset.

9.4.2 EASA's Four Ethical Pillars for Trustworthy AI

According to *EASA (2022)*, trustworthy AI in aviation rests on four normative pillars:

1. Respect for Human Autonomy. Humans must retain meaningful control over all AI-supported decisions. Automation may advise or act, but responsibility remains human. This principle underpins *human-in-the-loop* and *human-on-the-loop* architectures adopted for eVTOL flight control (EASA, 2024b).
2. Prevention of Harm. AI should enhance safety and environmental sustainability while minimizing unintended negative consequences. In AAM, this includes ensuring that algorithmic optimization does not compromise passenger welfare, data security, or environmental integrity (EASA, 2022, p. 10).
3. Fairness. Access to AAM must not deepen socioeconomic divides. Algorithms governing pricing, route allocation, or emergency prioritization must avoid discriminatory bias and be auditable for equitable outcomes (Floridi et al., 2021).
4. Explicability. Stakeholders—including regulators and passengers—must be able to understand why an AI system reached a specific conclusion. This demand for *explainable AI (XAI)* transforms transparency from an ethical aspiration into a certification criterion (EASA, 2024b).

Together, these pillars create a trustworthiness framework linking technical verification with moral accountability. On page 9 of *EASA (2022)*, the agency visualizes this relationship as concentric layers of governance—from ethical principles at the core to safety oversight and societal impact at the outer ring. This systemic model redefines ethics as a *governance continuum*, not a discrete checklist.

9.4.3 Embedding Ethics into Design and Certification

Translating these principles into operational practice requires integrating ethics into every phase of AAM development. EASA (2024b) proposes an Ethics-by-Design process combining system-engineering standards (e.g., DO-178C for software assurance) with qualitative assessments of human factors. This methodology ensures that explainability, accountability, and bias mitigation are validated alongside aerodynamic and structural tests. Airbus's *Project Vertex* and Joby Aviation's *Autonomous Mission Management AI* exemplify this convergence: both employ model-based reasoning that can generate human-readable justifications for autonomous actions (Airbus, 2023; Joby, 2024).

The ethical dimension also influences data governance. AI training for eVTOL operations depends on large sensor and flight datasets, raising issues of consent and surveillance. The *General Data Protection Regulation (GDPR)* and ICAO's *Manual on Cybersecurity (2022)* jointly mandate data minimization and contextual purpose limitation. Aligning aviation data frameworks with these civil-rights instruments strengthens societal legitimacy and harmonizes global practice.

9.4.4 Public Engagement and Societal Acceptance

Ethics in AAM cannot be confined to engineering laboratories or certification offices; it must engage the communities affected by aerial operations (Table 9.3). Studies by the *National Academies of Sciences (2024)* and *OECD (2024)* demonstrate that participatory governance—public hearings, online noise-mapping portals, and transparent environmental dashboards—reduces resistance to vertiport construction by up to 40%. Paris's *Réinventer Paris Air Mobility* project applied this model, combining citizen consultation with disclosure of algorithmic noise-routing parameters (ADP Group, 2024). In Dubai, the General Civil Aviation Authority (GCAA) integrates community feedback into its *AAM Regulatory Sandbox*, using sentiment analysis to adjust operational limits.

From a human-centric perspective, the ethical question is not merely whether AAM can operate safely but whether it operates justly—balancing innovation with respect for local identity, privacy, and quality of life. The *World Economic Forum (2025)* frames this as "ethical resilience," a city's capacity to absorb technological change without eroding civic trust. Ethical governance thus becomes both a moral imperative and a precondition for sustainable market entry.

9.4.5 Toward a Human-Centric Ethical Framework

Ethics in AAM must be conceived as a living system—dynamic, participatory, and adaptive to contextual values. The EASA model's emphasis on explicability and fairness demonstrates that trust cannot be programmed; it must be cultivated through openness and reciprocal learning between technology developers and society. Integrating these principles into global regulation requires a polycentric approach where international standards provide a moral baseline, but local cultures refine their implementation.

Ultimately, ethical assurance should deliver three outcomes:

1. Citizens understand how AI decisions affect them;
2. Operators remain accountable for automated actions; and
3. Regulators possess auditable evidence of fairness and transparency.

TABLE 9.3
Comparative ethical and public-engagement approaches in global AAM ecosystems

Region/Framework	Ethical Foundations	Public-Engagement Mechanism	Operational Integration	Key Reference
European Union (EASA)	Four Pillars: Human Autonomy, Prevention of Harm, Fairness, Explicability	Citizen assemblies, open-data noise maps	Ethics-by-Design integrated with type certification	*EASA (2022); EASA (2024b)*
United States (FAA/NASA)	Human oversight and accountability focus	Stakeholder forums through NASA AAM Community Integration Working Group	Voluntary Ethical AI Guidelines and FAA Innovate28 public comment cycles	*FAA (2023a); NASA (2024)*
Japan (METI/MLIT)	Fairness and privacy under Society 5.0 vision	Local prefectural consultations and social acceptance panels	Human-on-the-loop mandate for autonomous flights	*METI (2024); MLIT (2024)*
UAE (GCAA)	Ethical resilience linked to national ESG strategy	Digital sentiment surveys and public sandbox demonstrations	Integration of AI ethics into licensing and operator training	*GCAA (2024)*

When these conditions align, AAM transitions from a novelty to a socially embedded mobility system—one that mirrors society's highest aspirations for technological progress anchored in human dignity.

9.5 LONG-TERM POLICY TRENDS IN AAM REGULATION

Dimitrios Ziakkas, Eva Maleviti, and Fay Roberta Tsentemeidou

9.5.1 A Comparative Lens on Policy Innovation

The rapid maturation of the AAM sector has catalyzed diverse regulatory innovations that reveal how nations translate sustainability, cybersecurity, and ethical governance into actionable frameworks. While technological readiness levels (TRLs) for eVTOL platforms are converging globally, the *policy readiness level (PRL)* varies widely. Comparative assessment of the European Union, the United States, Japan, and the United Arab Emirates demonstrates that sustainable AAM integration is not merely a technical process but a *governance ecosystem* that balances public trust, digital safety, and environmental stewardship (EASA, 2024b; FAA, 2023b; METI, 2024).

In the European Union, EASA's Innovative Air Mobility (IAM) Framework and AI Roadmap 2.0 fuse sustainability and ethics into certification, establishing Europe as the first jurisdiction to require life cycle carbon accounting and explainable AI verification during type approval (EASA, 2024b). In the United States, the FAA Innovate28 Plan *and* NASA's National Campaign III take a sandbox approach: operations are authorized through phased data collection and public reporting rather than prescriptive metrics. This fosters industrial flexibility but risks uneven sustainability oversight. Japan's METI and MLIT have adopted a *Society 5.0* ethos—framing AAM as a socioeconomic infrastructure to enhance regional equity. In contrast, the UAE's GCAA situates AAM within its Green Logistics Strategy 2030, using ethical and ESG indicators as prerequisites for operator licensing.

9.5.2 Institutionalizing Sustainability across Domains

Global practice shows that sustainability in AAM is moving from aspiration to regulation. EASA integrates life cycle-emission analysis (LCA) directly into CS-VTOL.500, requiring manufacturers to quantify energy consumption per passenger-kilometer and noise-exposure maps before certification (EASA, 2023). FAA's *Environmental Policy Order 1050.1F* similarly mandates that vertiport projects undergo carbon and acoustic impact reviews. Japan's METI links sustainability to industrial policy by subsidizing hydrogen-hybrid propulsion research under the *Green Innovation Fund 2024*. The UAE extends this principle to governance, tying AAM development to national *Sustainable Aviation Fuel (SAF)* adoption targets and the UAE Net-Zero 2050 Strategy (GCAA, 2024).

Cross-analysis of these policies suggests a convergence toward integrated sustainability certification, where energy, environmental, and ethical audits coexist. This multi-pillar approach transforms sustainability from a single metric into a dynamic interaction of *carbon neutrality, social fairness, and digital responsibility.*

9.5.3 Harmonizing Cybersecurity and Ethical Oversight

A defining innovation of recent years is the harmonization of cybersecurity and ethics within regulatory regimes. The EU's Regulation (EU) 2023/203 obliges aviation organizations to maintain cybersecurity management systems aligned with ISO/IEC 27001 and ethical AI oversight committees (EASA, 2024b). The FAA's *AC 119–1* and the *Cybersecurity Strategy 2023* emphasize resilience, while Japan's MLIT framework mandates redundancy and continuous human supervision for autonomous operations (MLIT, 2024).

The UAE has moved further by establishing an AI Ethics and Safety Unit within its GCAA that cooperates with academic partners from Khalifa University to audit algorithmic bias and explainability (GCAA, 2024). Each of these initiatives converges on the principle that trustworthiness equals interoperability—technical, ethical, and cultural. As global eVTOL networks expand, interoperability of certification standards becomes a prerequisite for cross-border operations. ICAO's *Aviation Cybersecurity Strategy 2020–2030* thus acts as the normative bridge for these national efforts (ICAO, 2023a).

9.5.4 The Figure of Integration: A Human-Centric Model

The relationships among environmental, digital, ethical, and governance domains can be conceptualized as a *socio-technical ecosystem*. The following diagram (Figure 9.1) synthesizes the insights from EASA (2022, 2024b), FAA (2023a), METI (2024), and GCAA (2024). It illustrates how sustainability in AAM depends on the reciprocal reinforcement of four pillars—Environmental Sustainability, Cybersecurity & AI Safety, Ethical Governance, and Policy & Public Trust—all centered on the Human-Centric AAM System.

9.5.5 Insights from the Comparative Case

This comparative analysis highlights several cross-jurisdictional insights:

1. Integrated governance delivers resilience. Regions that embed environmental, cybersecurity, and ethical oversight within a single institutional architecture—such as EASA's IAM—demonstrate faster regulatory adaptation and higher public acceptance.
2. Transparency accelerates adoption. Both FAA and GCAA rely on public dashboards and stakeholder feedback to legitimize new routes; this participatory transparency reduces resistance and facilitates commercial viability.
3. Cultural context matters. Japan's framing of AAM within *Society 5.0* treats technology as a social equalizer, ensuring that innovation supports demographic and regional cohesion rather than privilege.
4. Economic alignment enhances continuity. The EU and UAE link AAM certification with ESG-oriented financing, anchoring sustainability in economic incentives rather than moral persuasion alone.

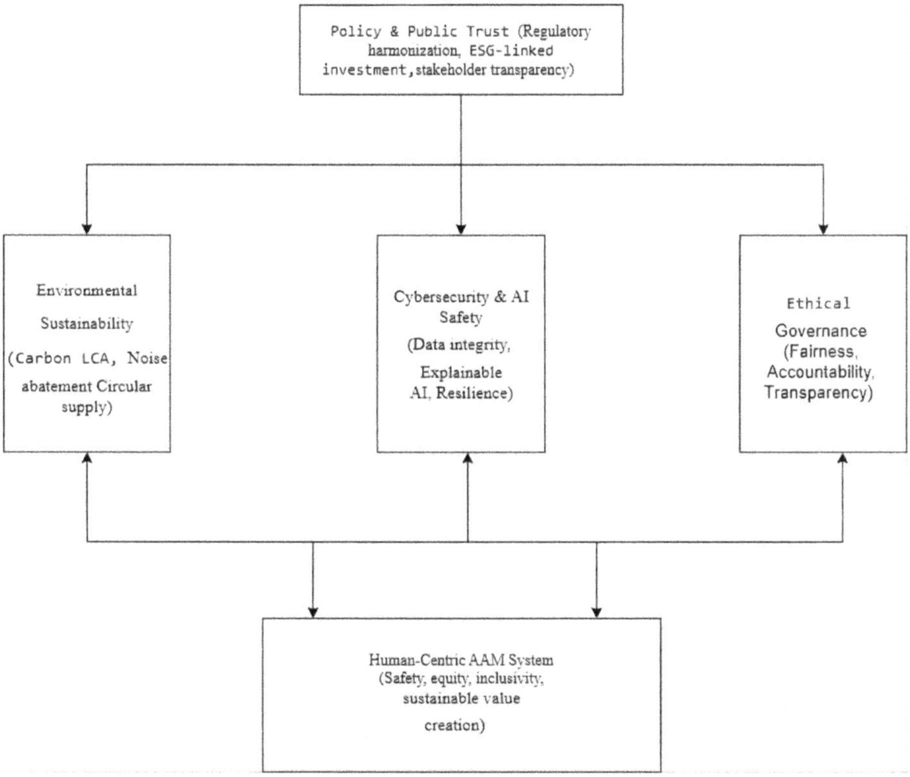

FIGURE 9.1 Framework for sustainable integration of AAM ecosystems.

Source: Synthesized by author based on EASA (2022, 2024b), FAA (2023b), METI (2024), and GCAA (2024).

9.6 CONCLUSION

Dimitrios Ziakkas, Aspasia Argyrou, and Debra Henneberry

The evolution of AAM represents one of the most consequential transformations in the history of civil aviation. Yet its promise will not be measured by how quickly aircraft take to the skies but be measured by how responsibly they integrate into the social, digital, and ecological fabric of human life. Sustainability, cybersecurity, and ethics form the triad of legitimacy upon which this transformation depends. Each dimension reinforces the others: without environmental integrity, the system loses moral justification; without digital resilience, it loses operational trust; and without ethical governance, it forfeits societal acceptance.

From an environmental perspective, AAM has the capacity to redefine aviation's carbon narrative. By coupling electrification with renewable-energy grids and closed-loop battery economies, eVTOL systems can displace the most carbon-intensive short-haul operations. Yet sustainability is not a technical endpoint—it is a social contract between innovators and the communities that host their infrastructure. Public confidence grows when citizens perceive fairness in access, transparency in noise management, and genuine participation in decision-making. These human-centered factors transform environmental metrics into ethical experiences of progress.

Cybersecurity, equally, is no longer an invisible technical layer but the digital expression of safety culture. In a networked AAM ecosystem, resilience depends on human vigilance as much as encryption strength. Regulators worldwide now recognize that cybersecurity assurance must be embedded within the same governance frameworks that manage safety and ethics. The alignment of *EASA (2024b)*, *FAA Innovate28*, *METI's AAM Roadmap 2040*, and *GCAA's AAM Safety Framework* signals a new maturity: security is being treated not as a defensive cost but as an enabler of public trust.

Ethical governance remains the integrating force among these domains. The *EASA (2022)* Trustworthiness Framework and similar initiatives by ICAO and NASA mark a shift from compliance-based oversight to value-based regulation. Human autonomy, fairness, explicability, and the prevention of harm are now formal design principles, ensuring that AAM technology evolves under the guardianship of human dignity. In this context, ethics functions as a unifying architecture that translates environmental and cyber resilience into social legitimacy.

Taken together, these insights reveal AAM as a microcosm of the 21st-century sustainability challenge: innovation must reconcile efficiency with empathy, precision with accountability, and automation with human purpose. The convergence of environmental stewardship, digital trust, and ethical integrity defines the threshold between technological novelty and transformative societal advancement. AAM thus becomes not merely an aviation revolution but a rehearsal for a more balanced relationship between humans, machines, and the planet they share.

Looking forward, the task for policymakers, engineers, and communities is to sustain this balance through continuous learning. The future of AAM will depend less on breakthroughs in propulsion or autonomy than on the collective ability to preserve trust as infrastructure—a renewable resource generated through transparency, inclusivity, and respect. When trust, sustainability, and ethics coalesce into the DNA of mobility, AAM can truly fulfill its promise: flight that serves humanity, not merely technology.

REFERENCES

ADP Group. (2024). *Reinventing Air Mobility: Noise impact assessment for Paris Olympics demonstrator*. Aéroports de Paris. https://doi.org/10.48364/adp.aam.2024

Airbus. (2023, November 22). *At your fingertips: Airbus flies a fully automated helicopter with a tablet*. https://www.airbus.com/en/newsroom/press-releases/2023-11-at-your-fingertips-airbus-flies-a-fully-automated-helicopter-with-a-tablet

Bekele, T., Kim, H., & Park, S. (2023). Machine-learning assurance for autonomous flight control systems. *Aerospace Science and Technology*, *141*, 108050. https://doi.org/10.1016/j.ast.2023.108050

DOE. (2024). *Domestic manufacturing for electric aviation components initiative*. U.S. Department of Energy. https://doi.org/10.2172/doe.ea.2024

EASA. (2022). *Ethics for artificial intelligence in aviation*. European Union Aviation Safety Agency. https://doi.org/10.2822/702911

EASA. (2023). *Certification specifications for Vertical Take-Off and Landing (VTOL) aircraft—Subpart E*. European Union Aviation Safety Agency. https://doi.org/10.2822/98877

EASA. (2024a). *Innovative Air Mobility implementation framework*. European Union Aviation Safety Agency. https://doi.org/10.2822/99712

EASA. (2024b). *Artificial Intelligence roadmap 2.0*. European Union Aviation Safety Agency. https://doi.org/10.2822/103978

ENISA. (2023). *Aviation cybersecurity threat landscape 2023*. European Union Agency for Cybersecurity. https://doi.org/10.2824/enisa.2023.aviation

FAA. (2023a). *Advanced Air Mobility (AAM) implementation plan—Innovate28*. Federal Aviation Administration. https://www.faa.gov/AAM-I28

FAA. (2023b). *Advisory circular AC 119–1: Aircraft systems information security protection*. Federal Aviation Administration. https://www.faa.gov/documentLibrary/media/Advisory_Circular/AC_119-1.pdf

Floridi, L., Cowie, C., & Beltrametti, M. (2021). How to design AI for social good: Seven principles for ethics by design. *Nature Machine Intelligence, 3*(3), 94–97. https://doi.org/10.1038/s42256-021-00356-y

GACA. (2024). *UAE green logistics strategy 2030 and AAM safety framework*. General Civil Aviation Authority. https://doi.org/10.56864/gaca.2030

IATA. (2023). *Guidance for aircraft battery recycling and second-life applications*. International Air Transport Association. https://doi.org/10.47468/iata.batt.2023

ICAO. (2023a). *Aviation cybersecurity strategy 2020–2030 and AAM study group report*. International Civil Aviation Organization. https://doi.org/10.55837/icao.cyber.2023

ICAO. (2023b). *Environmental report 2022/2023*. International Civil Aviation Organization. https://doi.org/10.55837/icao.env.2023

ICCT. (2024). *Lifecycle greenhouse-gas emissions from urban air mobility aircraft*. International Council on Clean Transportation. https://doi.org/10.48190/icct.uam.2024

Joby Aviation. (2024). *Autonomous mission management AI technical white paper*. Joby Aviation Inc. https://doi.org/10.55673/joby.ai.2024

Lee, D. S., Fahey, D. W., Skowron, A., et al. (2021). The contribution of global aviation to climate forcing for 2000–2018. *Atmospheric Environment*, *244*, 117834. https://doi.org/10.1016/j.atmosenv.2020.117834

METI. (2024b). *Japanese Advanced Air Mobility roadmap 2040*. Japan: Ministry of Economy, Trade and Industry.

MLIT. (2024a). *Guideline for secure operation of unmanned aircraft and AAM systems*. Japan: Ministry of Land, Infrastructure, Transport and Tourism.

NASA. (2024). *Advanced Air Mobility mission integration plan*. National Aeronautics and Space Administration. https://doi.org/10.25967/NASA.AAM.2024

NASEM. (2024). *Advanced Air Mobility and community outreach: A primer for successful stakeholder engagement*. National Academies Press. https://doi.org/10.17226/27627

NREL. (2023). *FAA vertiport electrical infrastructure study*. National Renewable Energy Laboratory. https://doi.org/10.2172/86245

OECD. (2024). *Urban mobility outlook 2024: Economic and environmental transitions*. OECD Publishing. https://doi.org/10.1787/umo-2024-en

Stansfeld, S. A., & Clark, C. (2015). Health effects of noise exposure in children. *Current Environmental Health Reports*, *2*(2), 171–178. https://doi.org/10.1007/s40572-015-0044-1

World Economic Forum. (2025). *Global aviation sustainability outlook 2025*. World Economic Forum. https://reports.weforum.org/docs/WEF_Global_Aviation_Sustainability_Outlook_2025.pdf

Zhao, Y., Sun, Q., & Li, B. (2023). Life-cycle energy analysis of electric aviation batteries. *Energy*, *274*, 127428. https://doi.org/10.1016/j.energy.2023.127428

10 Future Outlook and Global ROI Analysis of AAM

Dimitrios Ziakkas, Amin Vafadar, and Vasileios Vasileiou

10.1 INTRODUCTION

Advanced Air Mobility (AAM) represents a structural economic transition within the global transportation landscape, projected to evolve into a USD 1 trillion mobility ecosystem by 2040 (Morgan Stanley, 2024). The rapid emergence of electric Vertical Take-Off and Landing (eVTOL) vehicles and associated infrastructure—vertiports, digital traffic systems, and hybrid energy grids—marks a profound shift from fossil-fuel-based air transport toward electrified, distributed, and human-centric mobility systems. This shift aligns with broader sustainability frameworks, including the United Nations' *Sustainable Development Goal 9 (Industry, Innovation, and Infrastructure)* and the *European Green Deal's 2050 neutrality target*. From an investment standpoint, AAM promises multilayered returns on investment (ROI) encompassing not only direct financial yields but also societal, environmental, and productivity dividends.

The macroeconomic potential of AAM derives from three converging dynamics: (a) electrification of transport and carbon-neutral policy mandates; (b) digital integration across urban systems; and (c) increasing congestion costs in megacities. According to the *OECD Urban Mobility Outlook (2024)*, global urban congestion is estimated to impose annual economic losses exceeding USD 900 billion, with a compounded 5% annual increase in infrastructure strain. In this context, eVTOL networks offer regional governments an avenue to recapture lost productivity, expand intermodal connectivity, and stimulate high-value employment in the aerospace and renewable energy sectors. Chapter 10 provides an overview of the future outlook and return of investment analysis of the implementation of AAM in a global context.

10.2 GLOBAL ROI FORECASTS: URBAN AND REGIONAL APPLICATIONS

Dimitrios Ziakkas and Spiros Chazapis

Current ROI expectations for AAM infrastructure show differentiation between urban, suburban, and regional applications. Urban AAM—particularly airport

 DOI: 10.1201/9781003685555-10

shuttles and air-taxi networks—exhibits shorter payback horizons (6–10 years) but higher regulatory and public acceptance risks (EASA, 2024). Conversely, regional AAM—connecting peri-urban or remote zones—demonstrates more modest utilization rates but potentially stronger long-term yields through cargo, logistics, and medical operations. A balanced deployment across these domains is thus essential to realize sustainable growth.

- **Investment distribution and regional trends:** The regionalization of ROI underscores that AAM economics are context-sensitive (Table 10.1):

North America leads early-stage commercialization through companies like Joby Aviation and Archer, supported by the FAA's *Innovate28* plan. Market estimates project a total addressable market of USD 150–200 billion by 2035, with IRRs ranging 9–14% for integrated urban operations (FAA, 2023).

Europe's market, driven by EASA's *Innovative Air Mobility (IAM) Framework*, emphasizes sustainability and multimodal integration. The cumulative investment potential across the EU27 is expected to exceed EUR 120 billion by 2040, primarily focused on PPP-based vertiport networks and cross-border operations (EASA, 2024; SESAR, 2023).

Asia-Pacific, led by Japan, Singapore, and South Korea, exhibits high population density and advanced digital infrastructure, producing some of the strongest projected ROI for passenger eVTOL corridors, particularly in metropolitan areas like Tokyo-Yokohama and Seoul-Incheon (UAM Korea Roadmap, 2024). Japan's METI estimates a domestic eVTOL market of JPY 1.2 trillion by 2040 (≈USD 8 billion) (METI, 2024).

The Middle East, anchored by the UAE's *Dubai Urban Air Mobility Strategy*, positions AAM as part of its green logistics vision. Early modeling suggests AED 15 billion in cumulative ROI by 2035, underpinned by joint ventures with Volocopter and Skyports Infrastructure (GACA, 2024).

TABLE 10.1
Global ROI forecasts: urban and regional applications

Indicator	Quantitative Benchmark (2025–2040)	Primary Source
Direct employment multiplier	4–6 jobs per eVTOL aircraft produced	Guo et al. (2024a)
Payback period (urban AAM)	6–10 years	
Payback period (regional AAM)	8–14 years	EASA (2024)
Benefit-Cost Ratio (BCR)	1.3–2.1 (urban); 1.1–1.5 (regional)	Deloitte (2025)
Economic Impact Multiplier	1.6–2.3 (global mean)	Bridgelall (2023)
Noise reduction vs. helicopters	60–70% lower	EASA (2024)
CO_2 reduction (life cycle)	25–40% vs. conventional aircraft	

- **A systemic view of ROI:** Unlike traditional aviation investments, AAM's value is distributed across ecosystems, combining direct revenue streams (eVTOL operations, leasing, and vertiport management) and indirect multipliers (grid expansion, software services, and employment). The *Economic Impact Multiplier* for AAM is estimated to be between 1.6 and 2.3, indicating that each USD 1 invested generates up to USD 2.30 in wider economic output (Bridgelall, 2023). Moreover, the *Benefit-Cost Ratios (BCRs)* for early demonstrator cities—Los Angeles, Paris, and Dubai—range from 1.3 to 2.1, validating the feasibility of pilot deployments (Deloitte, 2025; OECD, 2024).

10.3 ECONOMIC AND SOCIAL BENEFITS OF AAM DEPLOYMENT

Volodymyr Bilotkach

The implementation of AAM operations from demonstration to daily service requires a place-sensitive approach: the infrastructure, operating concepts, and policy instruments that work in dense urban cores are not identical to those needed in suburbs or sparsely populated regions. This section outlines the key economic and social benefits that can result from deployment of AAM—including eVTOL passenger services, cargo drones, and vertiport infrastructure. In general, the deployment of eVTOL aircraft has the potential to reshape the global mobility landscape. In addition to creating new business opportunities, eVTOL operations will generate spillovers across industries, cities, and national economies. The benefits of eVTOL extend well beyond direct revenues from passenger and cargo services, encompassing job creation, new investment, productivity gains, enhanced regional competitiveness, and catalytic effects on related sectors. In these respects, the benefits of eVTOL business are similar to those of commercial aviation.

10.3.1 Economic Benefits

In this subsection, we will outline the key economic benefits associated with eVTOL deployment and development.

- **Job creation:** One of the most immediate and visible economic impacts of eVTOL operations is the creation of new jobs. Direct employment will arise in such areas as the design, development, and manufacturing of aircraft; day-to-day operations; maintenance; and the construction and management of vertiports. Each eVTOL entering service requires highly trained personnel for piloting, battery management, ground handling, and customer service. Vertiports, whether small rooftop pads or larger urban terminals, will generate additional roles in security, logistics, and passenger services.

Beyond these direct jobs, eVTOL deployment will stimulate indirect employment across its supply chains. Battery production, software development, construction, energy supply, and specialized materials manufacturing are all industries that will see growth as demand for eVTOL-related inputs increases. For example, the need for batteries will create opportunities for suppliers in the energy storage sector, while IT companies will develop fleet management solutions.

A further benefit is the creation of jobs in areas such as aerospace engineering, artificial intelligence, and digital services. Designing, certifying, and operating eVTOL systems require expertise that spans aerodynamics, avionics, electric propulsion, and AI-based scheduling algorithms. Over time, the accumulation of skills and knowledge around eVTOL technologies can help build advanced industrial clusters, positioning regions as global leaders in next-generation mobility.

Note that when evaluating job creation by a new industry, one should be mindful of making a distinction between job creation and job displacement. Some of the "new" jobs in the eVTOL sector might represent displacement from other sectors. For instance, to the extent development of eVTOL businesses affects such industries as helicopters or surface vehicle delivery, the workers might—to the extent their skills are transferable or following suitable reskilling—transfer to the eVTOL sector. In this case, the "new" eVTOL jobs may not represent genuine job creation at the national economy level.

- **New investment:** The emergence of eVTOL services promises to attract new investment into aerospace, mobility services, and supporting infrastructure. A considerable amount of venture capital money and private equity funding has already flowed into startups developing eVTOL prototypes, and this is expected to grow as certification milestones are reached. Governments and public agencies are also beginning to channel funding into vertiport construction, charging networks, and regulatory capacity, recognizing the potential economic returns.

These investments are associated with the multiplier effects. Local innovation ecosystems benefit as universities, research centers, and suppliers collaborate with eVTOL businesses on the development of new technologies. Startups focused on advanced materials, digital mobility platforms, or urban infrastructure design can use eVTOL growth as a springboard for scaling. The clustering of these activities around eVTOL hubs can stimulate regional innovation ecosystems similar to those seen in Toulouse and Seattle for the traditional aerospace industry. In this way, eVTOL operations act as a catalyst for broader industry growth.

- **Productivity and efficiency gains:** An important economic benefit of eVTOL can stem from productivity and efficiency improvements. By offering faster connections within and between cities, eVTOLs can deliver significant time savings for business travelers and commuters. Time saved this way translates directly into higher labor productivity, as workers spend less time in traffic. For businesses, faster travel also expands the effective labor market, enabling firms to access talent from a wider geographic area.

In the logistics sector, eVTOLs can improve supply chain speed and reliability. Rapid aerial deliveries of high-value or time-sensitive goods, such as medical supplies or just-in-time manufacturing components, reduce delays, and enhance supply chain resilience. The resulting improvements in productivity extend beyond individual firms to benefit the wider economy through higher competitiveness and economic growth.

- **Regional competitiveness:** Cities and regions that embrace eVTOL operations can stand to gain a competitive edge in attracting businesses. By integrating eVTOL services into their transportation ecosystems, metropolitan areas can alleviate road congestion, improve connectivity to airports, and provide a sustainable alternative to traditional transport modes. These enhancements make cities more attractive places to live, work, and invest.

For regions suffering from road congestion or a lack of surface transport options, eVTOLs can open up new opportunities by improving their access to economic hubs. Secondary cities and remote communities can be more closely integrated into regional economies, allowing businesses to operate more efficiently and residents to more quickly access services available throughout the region. Over time, this improved accessibility may support more balanced regional development, helping to reduce the concentration of economic activity in major urban cores. On the other hand, easier access to locations and services may also facilitate further concentration of businesses.

- **Catalytic effects on other sectors:** Further, eVTOL deployment can generate catalytic effects for other sectors of the economy. The tourism industry, for example, will benefit from easier access to attractions and destinations, encouraging longer stays and higher spending by (especially higher-income) visitors. Real estate markets around vertiports may experience growth, as proximity to fast and convenient aerial transport increases property values. Advanced manufacturing may also receive a boost, as the production of components for eVTOLs stimulates demand for precision engineering, materials science, and digital manufacturing technologies.

Finally, eVTOL operations will expand markets for clean energy and battery technologies. By requiring large-scale charging networks and high-performance energy storage solutions, the industry can help accelerate the development and diffusion of technologies critical to the global energy transition.

10.3.2 Social Benefits of eVTOL Operations

- **Enhanced mobility and accessibility:** Perhaps the most direct social benefit of eVTOL deployment lies in mobility enhancement. By operating above congested roads, eVTOLs—once they reach a certain scale—can reduce traffic density in urban centers. In addition to being a considerable inconvenience, congestion is also a source of lost productivity, elevated stress levels, and poor air quality. Introducing an aerial mobility layer may eventually redistribute traffic flows, offering commuters and travelers an alternative to ground journeys.

Importantly, eVTOLs offer a means of improving access to underserved or remote areas. Rural communities, islands, or regions with limited surface transport infrastructure can benefit from direct connections to economic hubs without the need for extensive infrastructure construction. This increased connectivity will allow the residents in such areas to access jobs, healthcare, and other services more easily. For urban commuters, eVTOL services can provide a faster, more reliable option for traveling across sprawling metropolitan regions, effectively shrinking cities and enabling greater flexibility in where people live and work.

- **Emergency and public services:** The societal value of eVTOL operations extends further into critical public service applications. Medical evacuations, organ transport, and the delivery of urgent medical supplies are all areas where eVTOLs and associated drone technologies can have life-saving impacts. In emergency situations, the ability to bypass congested roads and reach hospitals or disaster sites quickly can significantly improve survival rates and outcomes.

Disaster relief is another domain where eVTOLs can play a transformative role. Following floods, earthquakes, wildfires, or other natural or anthropogenic disasters, ground transport infrastructure is often damaged or inaccessible. In these situations, eVTOLs can provide rapid mobility for first responders, carry essential supplies, or facilitate evacuation. Faster response times not only save lives but also reduce the long-term social and economic costs of disasters by facilitating more effective recovery.

- **Environmental sustainability:** A further social benefit stems from the environmental sustainability of eVTOL technology. The eVTOLs rely on electric propulsion, which produces zero emissions at the point of use. When charged from renewable sources, eVTOL vehicles will also produce lower lifetime emissions. By replacing fossil fuel-based air and ground journeys, eVTOLs contribute to reducing greenhouse gas emissions, supporting decarbonization strategies at various levels—city, regional, national, and global.

Noise is another dimension of sustainability. Traditional helicopters are criticized for their noise footprint, which limits their acceptance in urban areas. eVTOL designs, with multiple small electric rotors, are substantially quieter, making them more compatible with dense environments. Lower noise and cleaner air enhance the quality of life in cities, improving both physical and mental health outcomes. These sustainability gains are not only an environmental achievement but also a social one, as they contribute to healthier, more livable urban environments.

- **Social inclusion:** While early eVTOL services are likely to be priced at a premium and targeted at high-income travelers, there is strong potential for greater social inclusion over time. As the industry scales and costs decline through advances in battery technology, economies of scale, and potential automation, prices per passenger-kilometer may correspondingly fall.

When this occurs, eVTOLs could democratize access to high-speed travel, no longer serving only premium customers but offering broader segments of society the opportunity to benefit.

In the longer term, integrating eVTOLs into public transport networks could ensure that aerial mobility is accessible to a wider range of users. This would represent not just a technological innovation but a meaningful advance in social equity, bringing the advantages of speed, accessibility, and connectivity to people across income groups and geographies.

10.4 INFRASTRUCTURE INVESTMENT AND LONG-TERM COST SAVINGS

Volodymyr Bilotkach

10.4.1 Considerations for Infrastructure Investment

The rollout of AAM, and particularly the deployment of eVTOL aircraft, will require a new layer of infrastructure that blends aviation standards with urban mobility needs. Planning investment in this infrastructure is a complex issue. The relevant aspects of this planning include regulatory frameworks, technical requirements, financial viability, and community integration. Unlike traditional airports, AAM infrastructure must coexist with dense urban environments while also delivering safe, efficient, and scalable services. This underscores the need for a clear understanding of the considerations involved in ensuring that infrastructure investment supports not only the operational needs of AAM but also broader societal and environmental goals.

- **Regulatory and policy environment:** The regulatory and policy framework is a fundamental consideration in planning AAM infrastructure investment. Because eVTOL operations fall within the domain of aviation, infrastructure must comply with stringent safety standards established by authorities, most importantly, the regional or country's safety regulators.

Equally important is the alignment of infrastructure with emerging systems for airspace management. Unlike conventional aviation, AAM will operate at lower altitudes and in close proximity to urban environments, requiring the development of dedicated urban air mobility corridors and unmanned traffic management (UTM) systems. Infrastructure planning must therefore anticipate the digital and physical systems needed to enable safe and efficient operations in shared airspace, with the view of integrating these new airspace users into what might already be dense urban environments.

Local governance is also critical. Zoning laws, land use regulations, and environmental approvals will dictate where vertiports can be built and how they can operate. Municipalities will have to update planning codes to accommodate vertiport construction on rooftops or within existing transportation hubs while also addressing environmental assessments to ensure compliance with noise and air quality standards.

- **Site selection and land use:** Selecting appropriate sites for vertiports is a crucial step in building a viable AAM network. Proximity to demand centers such as business districts, airports, and major transport hubs will determine whether eVTOL services can offer meaningful time savings to passengers. Locations must also provide seamless integration with other modes of transport to maximize connectivity.

Land availability poses challenges in dense urban areas. Rooftop vertiports may offer a solution, but they require structural assessments, safety compliance, and efficient passenger access routes. On-the-ground vertiports, meanwhile, will have to compete with alternative land uses such as housing, commercial development, and green space.

Community acceptance is another critical factor. Even with lower noise profiles compared to helicopters, eVTOL operations may generate resistance if vertiports are located too close to residential neighborhoods. In general, we can expect NIMBY-ism to fuel resistance to vertiports, especially at the early stages of development of this new industry, when eVTOL business will likely cater primarily to the wealthier clientele.

- **Energy and charging infrastructure:** AAM operations will rely heavily on electric power, making energy and charging infrastructure a central planning consideration. Ensuring sufficient grid capacity and stability is essential to support frequent eVTOL flights, particularly during peak hours. Grid upgrades may be required in many urban areas, and coordination with energy utilities will be critical. At the same time, eVTOL industry may be able to piggyback here on the prospective grid upgrades that will be implemented to accommodate electric vehicles, as well as the energy-hungry infrastructure associated with the deployment of Large Language Models (LLMs).

Renewable energy integration is another area of opportunity. By incorporating solar panels, on-site energy storage, or partnerships with renewable energy providers, vertiports can reduce their environmental footprint while also improving resilience against grid disruptions. Such investments align with broader city decarbonization goals and can enhance the attractiveness of AAM to environmentally conscious stakeholders.

- **Integration with existing transport networks:** The success of AAM depends on its ability to complement rather than compete with existing transportation modes. Infrastructure planning should prioritize seamless connectivity with airports, metro and rail systems, taxis, and ride-hailing services. For passengers, the ability to book a multimodal trip will be central to the convenience of the service.

First- and last-mile solutions are especially important. Without efficient ground access to vertiports, much of the time-saving advantage of eVTOLs could be lost. Partnerships with ride-hailing platforms, autonomous shuttles, or public transit agencies can help ensure smooth door-to-door journeys. Surface road congestion may also adversely affect demand for eVTOL services. Infrastructure planning must account for this issue—vertiport access road enhancements might be needed to ensure the success of eVTOL business.

Cargo operations also require integration with logistics networks and warehousing facilities. By linking vertiports to distribution centers, eVTOL cargo services can provide rapid deliveries of goods, enhancing the efficiency of supply chains. Infrastructure investment should therefore account for both passenger and cargo needs, potentially within shared facilities.

- **Scalability and flexibility:** Demand uncertainty in this nascent industry necessitates infrastructure design with scalability in mind. Modular vertiport designs allow for phased expansion—this approach reduces initial capital expenditure while maintaining flexibility for future growth.

Infrastructure must also be future-proofed for anticipated technological shifts. The potential transition to autonomous eVTOL operations will alter requirements for crew facilities, air traffic management, and passenger handling. Likewise, as aircraft evolve to carry more passengers or cargo, vertiports must be adaptable to accommodate larger vehicle types. Designing with flexibility ensures that investments remain relevant over the long term and reduces the risk of obsolescence.

From a financial perspective, balancing capital intensity with uncertain demand is a central challenge. Vertiport construction, charging systems, and air traffic integration require significant upfront investment, but revenues will initially be limited as the industry scales gradually. Further, the demand level for eVTOL businesses is highly uncertain. To de-risk early projects, public-private partnerships (PPPs) may play a key role, with governments sharing the burden of initial investment in exchange for long-term economic and social benefits.

- **Safety, security, and resilience:** It goes without saying that infrastructure planning must prioritize safety, security, and resilience. Passenger and cargo screening procedures must be adapted to the scale of vertiport operations, ensuring security without creating bottlenecks. Cybersecurity is equally critical, as everywhere else in the aviation industry. Protecting these systems from disruption or attack is vital to maintaining public confidence.

Resilience to external shocks, such as grid failures or extreme weather events, must also be factored into design. Backup power systems, redundant communication networks, and weather-resilient structures will enhance operational continuity and reduce risks. By building safety and resilience into infrastructure from the outset, AAM operators can avoid costly retrofits and establish trust with regulators and the public.

10.4.2 Longer-Term Cost Savings from eVTOL Operations

- **Lower operating costs versus helicopters:** A core driver of eVTOL cost advantage relative to the helicopters is the fundamental difference in propulsion systems. Conventional helicopters rely on complex mechanical assemblies with numerous moving parts, each requiring intensive inspection, lubrication, and replacement. By contrast, eVTOL vehicles employ distributed electric propulsion systems that are mechanically simpler, with fewer moving components. This shift significantly reduces maintenance labor hours, parts replacement, and downtime, which collectively account for a substantial share of helicopter operating expenses.

In addition, electricity offers a marked advantage over aviation fuel. Not only is electricity generally cheaper per unit of energy, but its price is also less volatile than that of jet fuel or aviation gasoline. Over the long term, as renewable energy becomes more prevalent, operators may even secure preferential pricing by integrating with clean energy providers or using on-site generation and storage. Together, these advantages suggest that eVTOLs could operate at a fraction of the per-hour cost of helicopters.

- **Infrastructure efficiency:** Traditional aviation infrastructure requires significant land and capital investment. By contrast, vertiports and charging hubs designed for eVTOL operations can be smaller, cheaper, and modular in nature. A rooftop vertiport may be integrated into an existing building, while ground-based facilities can be developed in underutilized urban plots. Modular designs allow phased investment, enabling operators and infrastructure providers to expand capacity incrementally in line with demand growth. This flexibility reduces upfront capital expenditures and lowers the risk of overbuilding.
- **Asset utilization:** Cost efficiency in aviation is heavily influenced by fleet utilization. eVTOLs offer several structural advantages here. Faster turnaround times, enabled by rapid battery charging and simplified ground handling, mean more flights per day per vehicle. In addition, distributed vertiport networks reduce deadhead repositioning flights, further increasing productive utilization. Over time, higher asset utilization drives down the cost per passenger-kilometer, allowing operators to spread fixed costs across a larger revenue base and achieve profitability at lower ticket prices.
- **Reduced road congestion:** Perhaps the most visible social cost saving could stem from reductions in road congestion. Urban traffic delays impose massive economic costs in the form of lost productivity, wasted fuel, and delayed goods. By shifting a portion of trips to the air, eVTOLs can relieve pressure on congested roadways, enabling smoother traffic flow for those who remain on the ground. Reduced congestion also slows the wear-and-tear on road infrastructure, lowering long-term government expenditures on repairs and expansions. It is too early to say whether eVTOL vehicles will be able to significantly reduce surface road congestion, as their uptake is at

this time subject to considerable uncertainty. One area that appears the most promising at this time is the removal of delivery vehicles from urban roads, as last-mile delivery will be increasingly handled by the eVTOL vehicles.

- **Reduction in road accidents:** Road accidents represent a significant economic and social burden. In addition to the cost of repairing or replacing the vehicles (and in some cases also the infrastructure), accidents lead to medical expenses, lost productivity, and human suffering. By substituting some road trips with eVTOL flights, cities may see reductions in accident rates and their associated costs.
- **Energy efficiency and emissions reduction:** Electric propulsion creates systemic savings through reduced dependence on fossil fuels and lower carbon emissions. While we can expect such impact from eVTOL vehicles to be significantly smaller than that from the surface electric vehicles, it will not be negligible either. By integrating with renewable energy systems, eVTOL infrastructure can further reduce the environmental footprint, aligning with decarbonization policies and avoiding future penalties associated with carbon pricing or emissions regulation.
- **Land use efficiency:** Finally, eVTOLs create long-term savings through more efficient land use. As urban populations grow, the pressure to expand road networks and parking facilities intensifies, requiring large capital investments and consuming valuable urban land. By shifting mobility to the air, eVTOLs reduce the need for these costly expansions. The freed land can instead be allocated to higher-value uses such as housing, green spaces, or commercial development, delivering both economic and social benefits to cities.

10.5 COMPARATIVE ANALYSES: eVTOL VERSUS TRADITIONAL TRANSPORTATION SYSTEMS

Dimitrios Ziakkas and Amin Vafadar

- **Economic and operational comparison:** Comparing eVTOL networks with traditional transportation systems reveals structural differences across cost structures, operational efficiency, and scalability (Table 10.2). The per-seat operating cost for electric VTOLs is estimated at USD 1.30–2.50 per passenger-kilometer, compared to USD 5–7 for conventional helicopter shuttle services (Chinthoju, 2025). eVTOLs also demonstrate 40–60% lower maintenance costs due to reduced mechanical complexity (Deloitte, 2025). However, initial CAPEX remains high, averaging USD 8–12 million per vertiport, depending on location and energy infrastructure (NREL, 2023).

From an operational perspective, eVTOL aircraft are optimized for short-to-medium-range (30–120 km) routes, while regional jets and high-speed rail dominate longer distances. Despite their limited range, eVTOLs outperform ground vehicles in congested corridors, achieving time savings of up to 70% per trip in metropolitan environments such as Los Angeles, Paris, and Singapore (SESAR, 2023).

- **Environmental and energy efficiency comparison:** eVTOLs offer significant environmental advantages over both traditional aviation and ground mobility. Life cycle emissions analyses conducted by the *International Council on Clean Transportation* (ICCT, 2024) indicate that eVTOLs generate 30–40% lower CO_2 emissions than helicopters per passenger-kilometer, even when accounting for electricity from mixed grids. When powered entirely by renewables, emissions can be reduced by up to 90%. Furthermore, energy efficiency reaches 110–150 Wh/passenger-km, substantially below that of turboprop aircraft (≈550 Wh/passenger-km) or internal combustion automobiles (≈500 Wh/passenger-km) (ICCT, 2024; NREL, 2023).
- **Societal and infrastructural implications:** While eVTOLs outperform traditional systems in efficiency and sustainability, scalability remains constrained by energy availability, certification complexity, and vertiport siting. Nevertheless, they promise a new spatial economics: linking peripheral regions to city centers without major infrastructure expansion. For instance, in Dubai, Volocopter's joint pilot project with GCAA demonstrated 45% shorter total journey times compared to car or metro routes, while achieving AED 0.35 per passenger-km in estimated operating cost once scaled (GACA, 2024). In Japan, SkyDrive's Osaka Bay corridor simulations project time savings of 60% and potential ROI of JPY 2.8 billion annually by 2035 (METI, 2024).

These comparative data underline that while traditional systems remain indispensable for mass transport, eVTOL networks fill the mobility gap—fast, electric, and point-to-point—between private cars and commercial aviation.

TABLE 10.2
Comparative performance metrics—eVTOL vs. traditional transport systems

Criterion	eVTOL (2025 est.)	Helicopter	Regional Jet	Electric Vehicle (EV)	High-Speed Rail (HSR)
Operating cost (USD/passenger-km)	1.30–2.50	5–7	0.12–0.20	0.09–0.12	0.08–0.10
Maintenance cost share of OPEX	12–15%	25–30%	20%	10%	8%
Average trip length (km)	30–120	100–300	300–900	15–60	300–800
Energy consumption (Wh/passenger-km)	110–150	400–600	550–700	180–220	90–120
CO_2 emissions (g/passenger-km)	25–40	80–100	70–85	50–70	25–30
Average ticket price (urban)	USD 50–120	USD 200–300	USD 60–100	USD 10–25	USD 30–60
Noise footprint (EPNdB)	65–70	90–95	80–85	65–70	70–75

Sources: Deloitte (2025); EASA (2024); ICCT (2024); NREL (2023).

10.6 LESSONS LEARNED AND BEST PRACTICES IN AAM ECONOMIC PLANNING

Dimitrios Ziakkas, Dimitra Synodinou, and Eleftherios Bokas

The transition toward a globally scalable AAM framework depends on institutional learning and strategic alignment between governments, industry, and communities. The lessons from early pilot programs reveal that economic viability and social acceptance depend on harmonizing technological innovation with policy integration and stakeholder engagement.

10.6.1 Public-Private Partnerships (PPPs) as Catalysts

PPPs have been the cornerstone of most successful AAM deployments. The Dubai Roads and Transport Authority (RTA) and Volocopter (Germany) established a vertically integrated partnership model that distributed financial and operational risks between public agencies and private developers (GACA, 2024). Similarly, Skyports and the UK Civil Aviation Authority (CAA) structured "Regulatory Sandbox" models enabling early testing of infrastructure economics under controlled conditions. PPPs ensure early-stage infrastructure financing—often covering 40–60% of CAPEX—while maintaining public oversight of safety and equity goals.

10.6.2 Integration of Energy and Digital Ecosystems

The success of AAM also depends on integration with energy grids and digital twins for predictive maintenance and airspace management (EASA, 2024). In Singapore, Skyports Infrastructure and Volocopter implemented a grid-interactive vertiport model, reducing power demand peaks by 25%. Integration of digital twins allowed real-time monitoring of battery degradation and load forecasting (NREL, 2023). These innovations enhance ROI by improving energy efficiency and minimizing downtime.

10.6.3 Regulatory Harmonization and Phased Certification

Regulatory clarity directly influences investment confidence. The FAA Innovate28 and EASA IAM frameworks employ phased certification—enabling progressive scaling from pilot routes to commercial corridors (EASA, 2024; FAA, 2023). Japan's Civil Aviation Bureau adopted a similar incremental pathway for SkyDrive, creating early ROI visibility for investors while preserving safety oversight (METI, 2024). Countries lacking clear certification processes—such as Brazil and India—face delayed investor participation despite technical readiness.

10.6.4 Social Acceptance and Stakeholder Engagement

Public trust remains an indispensable economic factor (Table 10.3). The *National Academies of Sciences* (2024) emphasize that acceptance of aerial mobility hinges on transparent communication, demonstrable safety performance, and equitable access. Paris' *Réinventer Paris Air Mobility* project showed that early community

TABLE 10.3
Global best practices and economic lessons from AAM demonstrator programs

Region/City	Program/Partnership	Financial Model	Key Economic Outcome	Lesson Learned
Dubai (UAE)	RTA-Volocopter Joint Initiative	PPP with risk-sharing	AED 15B projected ROI (2035)	Government leadership accelerates adoption
Paris (France)	Volocopter-ADP-EASA "Olympic Demo"	Public-private with EU funding	EUR 1.2B mobility infrastructure uplift	Transparency builds public trust
Osaka (Japan)	SkyDrive-ANA-Toyota	Corporate consortium	JPY 2.8B annual ROI (2035)	Industry alliances ensure scalability
Los Angeles (USA)	Joby-NASA-FAA Urban Air Mobility	Federal grants + VC funding	8–12% IRR over 10-year horizon	Federal coordination de-risks investment
Singapore	Skyports-CAAS Smart Vertiport	PPP + energy grid integration	25% reduction in grid demand costs	Digital-energy convergence increases ROI
Munich (Germany)	Lilium-Bavarian Gov't	Hybrid PPP	EUR 800M in regional tech investment	R&D clustering boosts economic spillovers

involvement—through environmental noise studies and transparent data sharing—reduced local opposition by 40% (EASA, 2024).

10.7 CONCLUSION

Dimitrios Ziakkas, Amin Vafadar, and Debra Henneberry

The global AAM sector stands at the intersection of mobility innovation and economic transformation. As governments seek sustainable transport solutions, eVTOL technologies offer not only operational efficiency but also systemic economic benefits. Yet, realizing AAM's potential depends less on technology and more on strategic governance, inclusive economics, and trust.

From a global ROI perspective, early projects validate positive benefit-cost dynamics, particularly in urban and peri-urban markets where congestion, emissions, and land scarcity drive demand. However, the transition from pilot programs to full scale networks demands resilient financial frameworks, continuous public-private collaboration, and harmonized certification pathways.

AAM represents a paradigm shift from infrastructure-heavy mobility to distributed aerial ecosystems. Its promise lies in fostering intermodal equality, where access to aerial mobility complements—not replaces—existing ground transport. For policymakers and investors, the guiding principle should be "sustainable scalability": ensuring that ROI integrates social welfare, carbon neutrality, and urban resilience.

Ultimately, AAM's future will be judged not only by its profit margins but also by its ability to enhance quality of life, inclusivity, and planetary stewardship—the hallmarks of truly human-centric innovation in the 21st century.

REFERENCES

Bridgelall, R. (2023). Forecasting market opportunities for urban and regional air mobility. *Technological Forecasting and Social Change*, *196*, 122835. https://doi.org/10.1016/j.techfore.2022.122835

Chinthoju, P. K. (2025). Cost per flight analysis of tilt-wing eVTOLs for urban mobility. *International Journal of Applied Engineering and Technology*, *6*(4), 110–120. https://doi.org/10.54634/ijaet.v6i4.159

Deloitte. (2025). *AAM economic viability and infrastructure outlook 2025*. Deloitte Insights.

EASA. (2024). *Innovative Air Mobility (IAM) Implementation framework*. European Union Aviation Safety Agency. https://doi.org/10.2822/99712

FAA. (2023). *Advanced Air Mobility (AAM) Implementation plan—Innovate28*. Federal Aviation Administration. https://www.faa.gov/AAM-I28

GACA. (2024). *Dubai Urban Air Mobility strategy annual report 2024*. UAE: General Civil Aviation Authority.

Guo, J., Chen, L., Li, L., Na, X., Vlacic, L., & Wang, F.-Y. (2024, May). Exploring the economic feasibility of advanced air mobility in the early stages. *IEEE Transactions on Intelligent Vehicles*, *9*(5), 4826–4830. https://doi.org/10.1109/TIV.2024.3422002

ICCT. (2024). *Lifecycle greenhouse gas emissions from urban air mobility aircraft*. International Council on Clean Transportation.

METI. (2024). *Japanese advanced air mobility roadmap 2040*. Japan: Ministry of Economy, Trade and Industry.

Morgan Stanley. (2024). *Future of mobility III: Electric and autonomous aviation.* Morgan Stanley Research.

NREL. (2023). *FAA vertiport electrical infrastructure study.* National Renewable Energy Laboratory. https://doi.org/10.2172/86245

OECD. (2024). *Urban mobility outlook 2024: Economic and environmental transitions.* OECD Publishing.

SESAR. (2023). *U-space urban air mobility demonstrator results report.* SESAR Joint Undertaking.

11 Global Case Studies in eVTOL Deployment and AAM Integration

Dimitrios Ziakkas, Debra Henneberry, and Amin Vafadar

11.1 INTRODUCTION

Advanced Air Mobility (AAM) has progressed from a speculative concept to a visible policy priority across major metros, driven by convergent pressures: urban congestion, the electrification of transport, and improvements in autonomy and digital airspace management. Yet the decisive test for AAM is not aircraft performance alone; it is contextual integration—how vehicles, vertiports, energy systems, digital infrastructures, and communities fit together in ways that are environmentally responsible, cyber-secure, ethically legitimate, and economically viable. The purpose of this chapter is to examine three emblematic deployments—Joby Aviation in Los Angeles (USA), Volocopter in Singapore, and Lilium in Munich (Germany)—and to distill comparative insights on return on investment (ROI) and public acceptance that can guide policymakers and operators as the sector moves toward scaled operations. Rather than offering exhaustive technicalities, the narrative foregrounds the human stakes of deployment: noise and privacy expectations, equity of access, transparency in algorithmic decisions, and the credibility of business models. These dimensions condition both regulatory trust and market demand. The European Union Aviation Safety Agency's study on societal acceptance emphasized that community concerns—especially *noise* and *safety*—shape willingness to accept flights and vertiports in dense urban fabrics, a finding echoed in industry and academic surveys globally (EASA, 2021).

A second anchoring theme is economic realism. Even optimistic forecasts acknowledge that near-term unit economics will depend on high utilization, carefully selected use cases (e.g., airport access), and measured scaling of infrastructure. Recent peer-reviewed analyses of AAM direct operating costs and fare ranges (with European case studies) underline that sustainable economics emerge from disciplined matching of vehicle capability, route length, load factor, and vertiport throughput—and that fares per passenger-km must fall toward the price band of ground taxis to unlock material mode share. These studies—grounded in transparent assumptions—provide more sober planning inputs than promotional materials alone (Pertz et al., 2023).

Finally, the case studies show that public acceptance and ROI are reciprocally reinforcing credible noise footprints, visible safety governance, and open engagement,

 DOI: 10.1201/9781003685555-11

all of which reduce the friction costs of siting vertiports and stabilizing demand. In turn, real usage enables price discovery and operational learning. Where policy ecosystems convene stakeholder's early—airports, transit agencies, grid operators, and community representatives—deployment curves steepen because trust is produced alongside infrastructure.

11.2 JOBY AVIATION IN LOS ANGELES, USA

Gary Vermaak, Dimitrios Ziakkas, and Amin Vafadar

- **Context and trajectory:** Los Angeles (LA) is a prototypical test bed: sprawling urban form, chronic congestion, and multiple region-serving airports. Joby Aviation has pursued an "airport access first" strategy, aligning with legacy aviation operations while preparing for city-pair missions. Milestones include FAA Part 135 Air Carrier Certification (2022) to operate on-demand services—initially with conventional aircraft—enabling procedural learning and customer-experience design before eVTOL type certification. In addition, the Delta-Joby partnership targets LA and New York as the first airport shuttle markets, integrating booking and passenger workflows. More recently, the FAA conducted human-in-the-loop simulations for eVTOL operations at LAX, de-risking procedures for a high-density hub (Joby Aviation, 2022).
- **Acoustics and community interface:** Public acceptance in LA is strongly tied to acoustic perception. NASA-Joby acoustic flight tests and subsequent communications indicate a far smaller sound footprint for Joby's aircraft than comparable rotorcraft under modeled LA routes. An independent trade-press summary of the modeling suggests that at cruise the area where the aircraft is louder than ambient can be orders of magnitude smaller than a helicopter on the same corridor, contingent on operational profiles and background noise. While not a substitute for certification-grade data in situ, these results establish a directionally credible narrative for quieter operations that can be explained to residents and airport-community forums (NASA & Joby, 2022).
- **Digital operations:** LA's pathway also involves software-enabled service orchestration (e.g., scheduling, passenger interfacing, and cockpit workload management). Regulatory progress on software approvals is material because urban operations demand granular time-and-space coordination among multiple flights and ground services. These steps are as much institutional integration as they are technical milestones (Reuters, 2024a).
- **Economic contours:** For ROI, airport shuttles leverage predictable demand and short door-to-gate transfer times for premium segments. Peer-reviewed cost models indicate that fares per km need to fall near ground taxi ranges in dense European contexts; in LA, similar logic applies, tempered by a higher value of time and congestion. Achieving this depends on a high load factor, vertiport cycle efficiency, and a progressive reduction in pilot costs as automation advances. In the near term, Joby's strategy—*earn operating know-how under Part 135, demonstrate acoustic acceptability, and phase*

in eVTOLs—is economically coherent for a hub-and-spoke use case centered on LAX and regional airports (Pertz et al., 2023).

- **Conclusion:** LA demonstrates the power of airport-anchored AAM as a first service. Where airport authorities, the FAA, and operators co-simulate procedures, communities receive tangible signals of safety and order, and investor narratives can rest on operational readiness rather than speculative scale curves (Joby Aviation, 2022).

11.3 VOLOCOPTER IN SINGAPORE

Dimitrios Ziakkas, Hans Cornelius Natakusuma,
Gary Vermaak, and Amin Vafadar

- **Context and trajectory:** Singapore presents a highly orchestrated urban system with strong regulatory capacity. In October 2019, Volocopter executed a crewed public flight over Marina Bay with support from Singaporean authorities, a demonstration designed to validate flight envelopes and gauge societal response. Subsequent Singapore Economic Development Board (EDB) communications signaled an ambition to launch services subject to multiagency approvals; however, more recent reporting indicates delays tied to local funding partnerships, illustrating the reality that AAM requires an ecosystemal commitment—vehicles, certification, vertiports, UTM, and capital alignment (Volocopter, 2019).
- **Urban airspace and planning:** Independent academic work has explored route-network design in Singapore's low-altitude airspace, proposing structured corridors and separation logic compatible with dense urban constraints—an early look at how city-specific airspace geometry could support regular services. Meanwhile, local research institutes have articulated autonomous UAM integration scenarios, connecting land use, equity, and policy to technology readiness. These efforts help transform a one-off demonstration into a governance blueprint for AAM as public infrastructure.
- **Public acceptance:** Singapore's consultative culture gives it an advantage in translating demonstrations into societal consensus. Comparative surveys (EU and industry) consistently place *noise, safety, and privacy* as top acceptance drivers; Singapore's compactness amplifies each of these concerns. Transparent noise dashboards, public trials, and participatory siting of vertiports are the credible pathways observed in other regions and applicable here (EASA, 2021).
- **Economics and staging:** The touristic Marina Bay loop and airport-city transfers remain the most viable near-term use cases in Singapore's geography. Economic analyses caution that utilization and network design dominate viability—short, scenic loops can prove profitable with touristic pricing, while daily commuting requires lower fares per km and higher throughput. The pause in commercial launch underscores that capital formation and ecosystem sequencing are as decisive as technology and regulation (Pertz et al., 2023).
- **Conclusion:** Singapore demonstrates that policy maturity and social license can move in lockstep with trials, but financing models and local partnerships

must be secured early to avoid momentum loss between demonstration and deployment (Volocopter, 2022).

11.4 LILIUM IN MUNICH, GERMANY

Dimitrios Ziakkas, Amin Vafadar, and Gary Vermaak

- **Context and trajectory:** Lilium's Munich-anchored thesis has been to develop regional air-mobility links that connect airports and economic clusters across Bavaria and neighboring states, leveraging airport safety cultures and longer stage lengths to improve unit economics. That framing distinguished the Lilium Jet—a ducted-fan, seven-seat architecture—from many short-hop urban eVTOL concepts and helped motivate early discussions with Munich and Nuremberg airports on network planning and infrastructure concepts (e.g., energy provisioning, passenger processing, and airside interfaces). Those 2021–2023 steps were coupled with regulatory progress: receipt of FAA G-1 certification as a basis for validation toward a dual EASA/FAA pathway and EASA Design Organization Approval (DOA) confirming organizational competence to design and hold a type certificate under SC-VTOL. These milestones placed Lilium on a credible certification path, at least on paper, anchored in Munich's airport ecosystem (AIN, 2025).
- **2025 inflection: insolvency and restructuring:** The picture changed materially in 2025. After a series of fundraising setbacks in late 2024, Lilium GmbH and Lilium eAircraft GmbH entered German insolvency proceedings, and by February 21, 2025, the successor operating entity, Lilium Aerospace, confirmed that operations had stopped while administrators sought rescue options. Public reporting describes an attempted €200 million investor consortium ("Mobile Uplift Corporation") that did not ultimately close; administrators then mandated KPMG to run an M&A process to sell core assets and IP. Throughout mid-to-late 2025, the Ambitious Air Mobility Group (AAMG) emerged as a prospective buyer for Lilium's assets; parallel reporting indicates some asset transfers already affecting Munich-area facilities (e.g., a specialized battery/laser-welding building at Oberpfaffenhofen being taken over by another Munich electric-aircraft firm, Vaeridion). As of November 2025, AAMG's acquisition bid remained under negotiation, while other buyers acquired portions of the portfolio, including patents. These developments mean that Lilium's going-concern status and program control are contingent on the outcome of insolvency proceedings rather than purely on certification progress (Aerotime, 2025).
- **Certification, orders, and the Munich thesis in context:** On the technical/regulatory side, previously communicated timelines had slipped to early 2025 for first piloted flights, with EASA type-certification and initial deliveries still publicly guided (pre-insolvency) for 2026. Lilium also publicized a large Saudi Arabian order (Saudia Group, up to 100 aircraft) in mid-2024—strategically significant for a regional-mobility thesis—though any commercial realization now depends on the restructuring's outcome. The dual-certification posture (EASA + FAA basis) remains an asset for any

successor that acquires the program, because it shortens the path to validation in multiple markets. However, capital adequacy and team continuity have become the binding constraints; the Munich airport-to-airport model is still operationally cogent, but program survival depends on securing ownership, talent, and facilities through the insolvency process (AIN, 2025).

- **Facilities and workforce implications in Bavaria:** The Oberpfaffenhofen facilities—important to Lilium's propulsion/battery industrialization and acoustics testing—illustrate how insolvency can ripple through regional industrial policy. Reporting indicates facility transfers and engineering attrition into neighboring firms, raising the ramp-back cost for any buyer and extending the timeline to restore Munich-based manufacturing at the prior scope. In practical terms, a future operator would need to reconstitute specialized capabilities (fireproof battery rooms, laser-welding systems, and acoustic halls) or contract them and rebuild DOA-qualified processes and supplier quality systems that insolvency typically disrupts. Those restart frictions have direct ROI consequences for the Munich regional network thesis because they delay certification flight tests, push out first revenue, and require fresh capital for reacquiring or replicating assets (Electrive, 2025).
- **Governance and market signaling:** For policymakers, the Munich/Lilium case now serves as a stress test of Europe's AAM industrial strategy: strong regulatory scaffolding (EASA SC-VTOL, DOA, G-1 cross-validation) can lower technical risk, but financial continuity and ecosystem-level commitments (airports, energy, and municipalities) are equally decisive. In a human-centric frame, community acceptance was aided by Lilium's emphasis on airport-to-airport routes that largely avoid dense residential overflight, a feature still valuable for any successor program. The lesson is not that the Munich model is flawed; rather, capital sequencing and asset stewardship must be resilient enough to carry promising programs through late-stage certification and industrialization—precisely when cash needs peak and patience is shortest (AWN, 2024).
- **Bottom line (as of November 2025):** Munich remains an attractive regional AAM node with sophisticated airport actors and a supportive aerospace labor market. But the Lilium Jet itself sits at an inflection: insolvency administration, active bids for assets/IP, partial facility transfers, and prior order announcements now coexist with still-valuable regulatory milestones. For readers of this chapter's comparative analysis, the case underscores that ROI and public acceptance hinge not only on technical merit and acoustic performance but also on institutional durability—the ability to finance, own, and steward critical assets through the final miles to certification and service launch.

11.5 COMPARATIVE ANALYSIS OF ROI AND PUBLIC ACCEPTANCE IN GLOBAL CASE STUDIES

Dimitrios Ziakkas, Aymane Chafik, and Vasileios Vasileiou

Across the three cases, economics and acceptance co-evolve with institutional cooperation:

11.5.1 Use Case Selection and Fare Realism

Peer-reviewed modeling indicates viable fare bands tightening toward ground taxi prices as utilization rises and cycle times fall. Airport shuttles (LA, Munich) exploit predictable demand and existing security processes; touristic loops (Singapore) can sustain premium pricing but are unlikely to unlock mass adoption without broader networks. These conclusions align with studies using transparent DOC breakdowns and sensitivity to load factor and flight cycles (Pertz et al., 2023).

11.5.2 Noise as an Economic Variable

Acoustic acceptability is not only a social good but also an economic input. Lower annoyance footprints reduce siting time for vertiports and constrain operational limitations less often, expanding revenue hours. NASA-industry testing on eVTOL acoustics and recent survey literature syntheses show active progress in prediction tools and community response research, both prerequisites for scalable ROI (NASA & Joby Aviation, 2022).

11.5.3 Digital Readiness and Airport Integration

Simulations at LAX and software approvals for service orchestration (booking, dispatch, and crew workload) are not peripheral—they unlock throughput, protect margins, and build regulator confidence. Airport-centric planning (Munich) embeds AAM into a mature safety culture, reducing start-up frictions and aligning with aviation governance (Joby Aviation, 2022).

11.5.4 Public Acceptance as Risk Capital

Comparative surveys (EASA, industry) confirm that transparent communication and co-design of routes/siting measurably improve acceptance (Table 11.1). Jurisdictions that publish noise footprints and engage residents early save both time and money (EASA, 2021).

11.6 LESSONS LEARNED FOR FUTURE AAM IMPLEMENTATION

Dimitrios Ziakkas, Ian Ibrook Tow Roth, and Gary Vermaak

- **Start where aviation already works:** The most robust early deployments attach to airports. This reduces certification ambiguity, leverages existing safety/security processes, and concentrates operations on corridors where communities already expect aviation activity. It also simplifies intermodal transfer design (rail, metro, and rideshare), making door-to-gate value propositions legible to the public (Joby Aviation, 2022).
- **Treat noise and transparency as infrastructure:** Acoustic performance must be explained, not merely asserted. Publishing modeled and measured noise hemispheres and providing auralizations for communities can transform abstract risk into informed judgment; this, in turn, shortens the queue of permits and hearings and improves capital efficiency (Riziq et al., 2024).

TABLE 11.1
Indicative comparative case studies overview

Dimension	Los Angeles—Joby	Singapore—Volocopter	Munich—Lilium
Primary use case	Airport access (LAX) + metro pairs	Touristic loops + airport access (paused)	Regional airport-to-airport
Institutional posture	FAA + LAX simulations; airline partnership channel	Multiagency orchestration; strong regulatory capacity; capital partnership gaps	Airport-operator partnerships; regional network planning
Acoustic narrative	NASA flight tests; small modeled footprint areas relative to helis (context-dependent)	Urban density drives stringent thresholds; demos build "sound familiarity"	Airport-to-airport reduces residential exposure
ROI levers	Utilization, software-enabled throughput, premium airport time value	Touristic pricing, need for ecosystem financing to scale	Longer stage lengths, airport competencies; business/corporate demand
Public acceptance	Framed via quieter operations + airport governance	Trust via state capacity; requires visible benefits + fair siting	Lower sensitivity due to airport alignment; transparency on regional routes

Sources: NASA/Joby tests and LAX simulation notes; EASA social acceptance study; MDPI cost and noise surveys; official operator and airport communications.

- **Sequence capital with ecosystem maturity:** Singapore's pause is instructive: a vehicle and demonstration are necessary but insufficient; vertiport, grid, UTM, financing, and workforce need synchronized commitments. Public-private roadmaps should include gates linking capital drawdowns to regulatory and community milestones (Volocopter, 2022).
- **Price discovery requires scale, but scale requires trust:** Studies show fare bands tightening with utilization; however, utilization rides on *trust.* Programs that invest in participatory siting, privacy-aware data governance, and visible ethical oversight (e.g., human-on-the-loop policies for autonomy) build the social license necessary to reach the load factors that make fares competitive (EASA, 2021).
- **Avoid one-size-fits-all architectures:** LA's urban shuttle, Singapore's demonstration-to-ecosystem path, and Munich's regional corridors show that mission design must fit geography, governance, and economic structure. Premature standardization risks misallocating capital into use cases that a city cannot absorb (Pertz et al., 2023).

11.7 CONCLUSION

Dimitrios Ziakkas and Debra Henneberry

The presented case studies converge on a simple proposition: AAM integration is a civic project disguised as an aerospace program. Technology enables the possibility of new connections, but *institutions, communities, and business models* determine whether those connections endure. In Los Angeles, airport-anchored operations and credible acoustic evidence are turning a congested region into a test bed for premium, time-saving services. In Singapore, regulatory competence and urban-planning rigor promise high standards—if ecosystem financing is staged to match ambition. In Munich, regional networks harness airport governance to bend economics in favor of longer-leg utility before democratization.

Across all three, *ROI and public acceptance* are two sides of one ledger: where trust is earned—through quieter operations, transparent procedures, and accountable automation—capital finds confidence, permitting the scale that lowers fares and broadens access. Conversely, where trust is deferred, costs compound and horizons recede. The near-term challenge is thus not only solving remaining certification hurdles but also governing well: align noise disclosure with community dialogue; make software and airspace simulations public; stage finance with transparent milestones; and explicitly measure equity in access. Done this way, AAM can mature as an *ethically grounded, economically credible* addition to the world's mobility ecosystems.

REFERENCES

AeroTime. (2025, February 21). *Lilium halts operations and files for insolvency in Germany.* https://www.aerotime.aero/articles/lilium-halts-operations-and-files-for-insolvency

AIN Online. (2025, September 15). *Ambitious Air Mobility Group bids for Lilium assets under insolvency administration.* https://www.ainonline.com/aviation-news/advanced-air-mobility/2025-09-15/aamg-bid-lilium

Aviation Week Network. (2024, October 3). *Lilium delays manned flight to early 2025, seeks additional funding.* https://aviationweek.com/advanced-air-mobility/lilium-delays-manned-flight-early-2025

EASA. (2021). *Study on the societal acceptance of Urban Air Mobility in Europe.* European Union Aviation Safety Agency. https://www.easa.europa.eu/sites/default/files/dfu/uam-full-report.pdf

Electrive. (2025, July 11). *Vaeridion takes over Lilium's Oberpfaffenhofen facility as asset sale progresses.* https://www.electrive.com/2025/07/11/vaeridion-takes-over-lilium-oberpfaffenhofen-facility

Joby Aviation. (2022, May 26). *Joby receives Part 135 Air Carrier Certificate from the FAA.* https://www.jobyaviation.com/news/joby-receives-part-135-air-carrier-certificate/

NASA & Joby Aviation. (2022). *Acoustic flight test of the Joby Aviation advanced air-mobility prototype vehicle (AIAA/CEAS 2022 Conference paper).* NASA Technical Reports Server. https://ntrs.nasa.gov/api/citations/20220006729/downloads/Aeroacoustics2022_Pascioni_STRIVES5.pdf

Pertz, J., Niklaß, M., Swaid, M., Gollnick, V., Kopera, S., Schunck, K., & Baur, S. (2023). Estimating the economic viability of advanced air mobility use cases: Towards the slope of enlightenment. *Drones*, *7*(2), 75. https://doi.org/10.3390/drones7020075

Reuters. (2024a, June 20). *FAA nod for Joby's in-house ElevateOS software for air-taxi operations.* https://www.reuters.com/business/aerospace-defense/joby-aviation-says-faa-gives-nod-in-house-software-air-taxi-operations-2024-06-20/

Reuters. (2024b, September 16). *Saudia Group to buy up to 100 Lilium Jets for regional mobility.* https://www.reuters.com/business/aerospace-defense/saudia-group-buy-up-100-lilium-jets-2024-09-16/

Riziq, M., et al. (2024). Noise prediction and mitigation for UAS and eVTOL aircraft: A survey. *Drones*, *9*(8), 577. https://doi.org/10.3390/drones9080577

Volocopter. (2019, October 22). *Volocopter air taxi flies over Singapore's Marina Bay*. https://www.volocopter.com/en/newsroom/volocopter-air-taxi-flies-over-singapores-marina-bay

Volocopter. (2022). *The launch of Urban Air Mobility in Singapore—A roadmap (White paper)*. https://assets.ctfassets.net/vnrac6vfvrab/7fevMxgxu1PQPekge6tnUQ/3bccc73083a23335a174442297661584/Volocopter_Whitepaper_Singapore-Roadmap_web__3_.pdf

12 Conclusions and Strategic Recommendations for Global AAM Implementation

Dimitrios Ziakkas and Aymane Chafik

12.1 INTRODUCTION

The development of Advanced Air Mobility (AAM) and electric Vertical Take-Off and Landing (eVTOL) systems is more than an engineering milestone—it represents a paradigm shift in how societies conceive movement, energy, and safety. The preceding chapters have examined the human, ethical, technological, and economic dimensions that define this transformation. The integration of digital autonomy, electric propulsion, and distributed traffic management creates a fundamentally interdisciplinary transportation ecosystem—one that requires the convergence of aerospace design, regulatory foresight, social sciences, and human-factors research.

In 2025, the global momentum toward AAM has accelerated. EASA's Innovative Air Mobility (IAM) Framework (2024a) and the FAA's Innovate28 Plan (2023a) have outlined near-term operational targets, while the UK Civil Aviation Authority (CAA) and UK Research and Innovation (UKRI) under the *Future Flight Challenge (2024)* are pursuing system-of-systems trials integrating hydrogen propulsion, AI-enabled air-traffic coordination, and public co-design. At the same time, NASA's AAM Mission Integration Plan (2024) and ICAO's forthcoming AAM Blueprint (2025) reflect a maturing international consensus: that safe, sustainable, and socially legitimate air mobility depends on human-centric implementation, not just technological readiness.

This concluding chapter consolidates these lessons and projects a strategic vision for the coming decade. It interprets the findings of earlier chapters—on sustainability, cybersecurity, ethics, economics, and public acceptance—through a single organizing lens: *trust*. Whether expressed as flight safety, certification credibility, or community acceptance, trust is the ultimate infrastructure of AAM. The sections that follow articulate the human and policy foundations needed to transform early trials into scalable, resilient, and equitable global networks.

DOI: 10.1201/9781003685555-12

12.2 KEY TAKEAWAYS ON TRAINING, OPERATIONS, AND HUMAN FACTORS

Dimitrios Ziakkas and Aymane Chafik

- **Human performance at the center of digital aviation:** Across regulatory and operational domains, human factors remain the critical determinant of AAM safety. While automation promises workload reduction, it introduces *cognitive asymmetry*: pilots, dispatchers, and maintenance engineers interact with opaque AI systems whose internal logic may be only partially transparent. Both EASA's AI Roadmap 2.0 (2024b) and the FAA Human Factors Design Guidance (AC 25-11B) emphasize the need for *explainable automation*—interfaces that reveal machine reasoning and support operator recovery from automation failure.

In the hybrid eVTOL cockpit, situational awareness becomes distributed between human and algorithm. Studies by NASA (2024) show that pilot performance in automated flight testing depends less on reaction time than on *trust calibration*: understanding when to rely on automation and when to reassert manual control. Consequently, training systems must evolve from task-based proficiency toward cognitive resilience—teaching pilots how to reason about system states, not merely how to manipulate controls.

- **Training ecosystem transformation:** Traditional type-rating paradigms are unsuited to eVTOLs that blend fixed-wing, rotary-wing, and fly-by-wire logic. EASA's eVTOL Flight Crew Licensing Concept Paper (2024a) proposes modular competency frameworks emphasizing scenario-based learning and *Evidence-Based Training (EBT)* principles. Similarly, the FAA's Integrated Pilot Training Demonstration (2023a) introduces *adaptive simulation loops* that integrate eye-tracking and physiological monitoring to quantify mental workload. These approaches are consistent with the human-centric paradigm advanced throughout this book: technology should adapt to people, not the reverse.

At an operational level, AAM demands new categories of expertise: *UAM flight dispatchers, vertiport safety managers, cybersecurity officers,* and *AI systems supervisors.* These roles blend aeronautical knowledge with data literacy and ethical awareness, requiring collaboration between aviation authorities and academic institutions. The UK Future Flight Challenge (UKRI, 2024) explicitly funds curriculum design linking universities, manufacturers, and regulators—a template for international emulation.

- **Operational culture and safety management:** AAM extends the boundaries of Safety Management Systems (SMS) into cyber-physical space. The EASA Cybersecurity Management System Regulation and the FAA's Advisory Circular AC 119-1 (2023b) require safety and cyber risk assessments to converge within the same organizational framework. Human factors here include not only pilot performance but also digital hygiene, human-machine interface trust, and error propagation through data systems.

Operational experience from early AAM testbeds—Los Angeles, Paris, Osaka, and Dubai—shows that Just Culture principles remain as relevant to software incidents as to mechanical failures. When algorithms err, human oversight must be protected from scapegoating, allowing transparent learning cycles. This reinforces the broader finding that safety culture is inseparable from ethical culture, and that human performance systems must extend to those who design, train, and audit AI within aviation.

12.3 POLICY RECOMMENDATIONS FOR eVTOL CERTIFICATION AND STANDARDIZATION

Dimitrios Ziakkas, Debra Henneberry, and Anastasios Plioutsias

- **Convergence without uniformity:** The coexistence of EASA's SC-VTOL framework, the FAA's Part 23 and Part 135 adaptations, and the UK CAA's CAP 2272 "Design Verification Pathways" (2024) underscores both innovation and fragmentation. While diversity allows experimentation, the absence of interoperable certification baselines threatens cross-border operations. The policy priority for 2025–2030 is therefore functional convergence: aligning safety objectives, data standards, and performance-based criteria without imposing rigid procedural uniformity.

ICAO's AAM Study Group (AAM-SG) is developing a *Global Airworthiness Alignment Matrix* to map equivalencies among SC-VTOL, Part 23/135, and national standards—a promising mechanism for mutual recognition. Governments should support ICAO's effort through transparent data sharing and coordinated validation campaigns, ensuring that the first certified eVTOLs can operate internationally without duplicative oversight.

- **From prescriptive to performance-based certification:** EASA's SC-VTOL explicitly moves away from prescriptive design rules toward performance-based certification (PBC), emphasizing demonstrated safety levels rather than predefined configurations. The FAA's *Means of Compliance (MoC)* framework mirrors this philosophy. Regulators should now institutionalize adaptive certification—a process allowing iterative software updates and continuous AI learning within certified systems, subject to monitored boundaries.

To manage these dynamics, authorities need regulatory sandboxes that balance innovation with public safety. The UK CAA Innovation Sandbox *and* EASA's AI Ethics Assessment Tool exemplify such mechanisms, permitting experimental data gathering under controlled risk envelopes. Policy should ensure that sandbox results are publicly documented to accelerate collective learning and avoid redundant experimentation.

- **Vertiport and infrastructure standardization:** Standardization must extend to ground infrastructure. ICAO Annex 14 Volume II revisions (in progress) and EASA's Vertiport Design Specifications (2024a) call for harmonized lighting, obstacle limitation surfaces, and energy interface standards. Governments should adopt open data schemas for vertiport location, capacity, and grid interconnection to facilitate national AAM integration planning. The European Commission's *U-space Regulation* provides a model for such digital infrastructure interoperability.
- **Ethics and accountability in AI-enabled certification:** Finally, regulators must integrate ethical accountability into certification. The EASA (2022) Ethics for AI in Aviation document defines *explicability* and *human oversight* as mandatory design values. Certification dossiers should include *AI audit trails*, explainability evidence, and bias-mitigation reports. This "ethics-by-documentation" complements traditional airworthiness evidence, ensuring that safety cases are not only technical but also moral statements of responsibility.

12.4 EMERGING TRENDS AND FUTURE RESEARCH DIRECTIONS IN AAM

Dimitrios Ziakkas, Amin Vafadar, and Debra Henneberry

- **Hybrid-energy propulsion and circular sustainability:** The decarbonization of aviation will accelerate through hybrid-energy architectures. Hydrogen-electric and hybrid-turbine eVTOL concepts under development in Japan, Germany, and the United Kingdom demonstrate potential for energy densities above 700 Wh/kg, surpassing current lithium-ion systems. Research must focus on safety certification of hydrogen storage, thermal management, and life cycle carbon accounting. The EU's *Clean Aviation Joint Undertaking (2024)* identifies hybrid systems as transitional enablers toward full-scale zero-emission flight by 2040.

Circular economy models—battery recycling, second-life applications, and responsible sourcing of critical minerals—require coordination between aviation and environmental agencies. The IATA (2023) and Environmental Certification Specifications already mandate carbon declarations for propulsion batteries, setting a precedent for global adoption.

- **Digital twins and predictive safety:** The convergence of big data, AI, and sensor fusion is transforming safety assurance. Digital-twin ecosystems allow continuous monitoring of aircraft and vertiport systems, predicting component failures and optimizing maintenance. NASA (2024) and the UK CAA (2024) have begun integrating digital twins into certification simulations, allowing virtual validation of design changes before physical modification. Future research should explore *human interpretability of twin outputs*—how operators can make sense of algorithmic prognostics without cognitive overload.

- **Societal research: acceptance, equity, and justice:** Social-science inquiry remains underfunded relative to engineering. Longitudinal studies are needed to track public acceptance, perceived fairness, and distributional impacts of AAM services. As highlighted by EASA's *Societal Acceptance Study (2021)*, acceptance rises when citizens perceive clear public value and ethical governance. Integrating participatory modeling—citizens co-designing noise maps, route networks, and pricing algorithms—could institutionalize legitimacy.
- **Integration of AI ethics and human-machine trust research:** AI explainability is both a technical and psychological frontier. Future research must unify cognitive ergonomics with computational transparency, designing interfaces that express algorithmic intent in human terms. Cross-disciplinary collaboration among computer scientists, psychologists, and human-factors specialists will be crucial. The EASA AI Trustworthiness Framework (2022) should evolve into a global research consortium linking academia, regulators, and industry.

12.5 STRATEGIC ROADMAP FOR GLOBAL eVTOL AND AAM ECOSYSTEM DEVELOPMENT

Dimitrios Ziakkas, Amin Vafadar, and Debra Henneberry

The strategic roadmap for global eVTOL and AAM ecosystem development follows a three-phase approach (Figure 12.1).

- **Phase I (2025–2030): Demonstration to integration/Near-Term Expansion**

The near-term horizon focuses on integrating pilot programs into metropolitan transport systems. By 2030, at least several cities—Los Angeles, Paris, Dubai, Osaka, Singapore, Munich, and Seoul—are expected to host limited commercial routes.

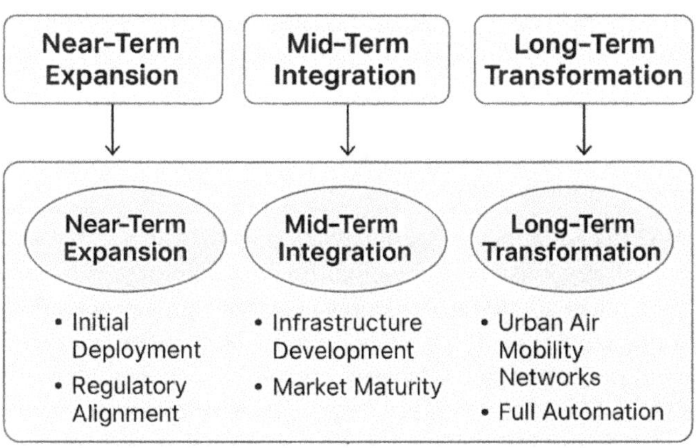

FIGURE 12.1 Strategic roadmap for Global eVTOL and AAM ecosystem development.

Key Strategic Imperatives Include:

- Institutionalizing *data-driven safety management* combining flight, maintenance, and cybersecurity metrics.
- Expanding *public-private partnerships (PPPs)* to share infrastructure costs and societal benefits.
- Establishing *cross-border certification reciprocity* through ICAO's Global Alignment Matrix.
- Embedding *ethical audits* within AAM operator licensing.
- **Phase II (2030–2040): network scaling and multimodal fusion/mid-term expansion:** By the 2030s, AAM will converge with autonomous ground and maritime mobility to form multimodal intelligent transportation grids. eVTOL networks will rely on standardized digital traffic management systems (UTM/U-space) integrated with national ATC. Governments should prioritize open-source interoperability, avoiding proprietary data monopolies that could fragment airspace management. Investment in renewable-powered vertiports and smart-grid integration will become a determinant of both environmental and financial sustainability.

Human capital will be decisive: an estimated 50,000–70,000 specialized AAM professionals will be required globally by 2040 (OECD, 2024). International training standards must evolve under ICAO and the International Society of Air Safety Investigators (ISASI) to ensure global competency recognition.

- **Phase III (2040 and beyond): autonomous, equitable, planetary/long-term expansion:** Long-term vision extends beyond technical autonomy to societal autonomy—communities empowered to co-govern aerial mobility as a public good. Fully autonomous eVTOLs, certified under continuous AI-assurance systems, will expand access to rural and underserved regions, promoting mobility equity. Ethical AI oversight will remain human in governance even as control becomes digital in operation. The global roadmap culminates not merely in fleets of electric aircraft but in a new social contract for airspace—transparent, sustainable, and just.

12.6 FINAL THOUGHTS: VISION FOR THE FUTURE OF ADVANCED AIR MOBILITY

Dimitrios Ziakkas, Amin Vafadar, and Debra Henneberry

The story of AAM is not about replacing aviation as we know it, but about rehumanizing flight through responsible innovation. Each technology examined throughout this volume—AI, electrification, autonomy—carries both promise and peril. The difference lies in governance. If societies treat AAM as infrastructure for collective benefit, grounded in transparency, fairness, and environmental stewardship, the sky can become a shared civic space rather than a privatized frontier.

The EASA, FAA, and UK CAA roadmaps collectively affirm that the aviation of the future will be data-driven yet human-centered, global yet locally accountable, and autonomous yet ethically governed. The trajectory from demonstration to normalization will test regulatory courage, social imagination, and interdisciplinary cooperation. Success will be measured not only in flight hours but also in public trust, carbon avoided, and communities connected.

In that vision, *Advanced Air Mobility* becomes the first truly *moral technology of flight*: one that flies not merely because it can, but because it should—safely, sustainably, and inclusively.

REFERENCES

CAA (UK Civil Aviation Authority). (2024). *CAP 2272—Design verification pathways to support innovation in aviation.* https://www.caa.co.uk/cap2272

EASA. (2021). *Study on the societal acceptance of Urban Air Mobility in Europe.* European Union Aviation Safety Agency. https://www.easa.europa.eu/sites/default/files/dfu/uam-full-report.pdf

EASA. (2022). *Ethics for Artificial Intelligence in aviation.* European Union Aviation Safety Agency. https://doi.org/10.2822/702911

EASA. (2024a). *Innovative Air Mobility (IAM) Implementation Framework.* European Union Aviation Safety Agency. https://doi.org/10.2822/99712

EASA. (2024b). *Artificial Intelligence roadmap 2.0.* European Union Aviation Safety Agency. https://doi.org/10.2822/103978

FAA. (2023a). *Advanced Air Mobility (AAM) Implementation plan—Innovate28.* Federal Aviation Administration. https://www.faa.gov/AAM-I28

FAA. (2023b). *Advisory Circular AC 119–1: Aircraft systems information security protection.* Federal Aviation Administration. https://www.faa.gov/documentLibrary/media/Advisory_Circular/AC_119-1.pdf

IATA. (2023). *Guidance for aircraft battery recycling and second-life applications.* International Air Transport Association. https://doi.org/10.47468/iata.batt.2023

NASA. (2024). *Advanced Air Mobility mission integration plan.* National Aeronautics and Space Administration. https://doi.org/10.25967/NASA.AAM.2024

OECD. (2024). *Urban mobility outlook 2024: Economic and environmental transitions.* OECD Publishing. https://doi.org/10.1787/umo-2024-en

UK CAA & UKRI. (2024). *Future flight: Integration of AAM into UK airspace—Joint policy paper.* Civil Aviation Authority.

World Economic Forum. (2025). *Global aviation sustainability outlook 2025.* World Economic Forum. https://reports.weforum.org/docs/WEF_Global_Aviation_Sustainability_Outlook_2025.pdf

Appendices

Appendix A: Regulatory References and Certification Requirements for eVTOL Integration

A.1 OVERVIEW

The certification of electric Vertical Take-Off and Landing (eVTOL) aircraft represents the first major regulatory convergence between the European Union Aviation Safety Agency (EASA) and the Federal Aviation Administration (FAA) since the introduction of composite-airframe and fly-by-wire standards in the 1990s. Both agencies recognize that eVTOLs, as hybrid rotorcraft-fixed-wing vehicles with varying levels of automation, demand a performance-based regulatory paradigm rather than a derivative of legacy Part 23 or CS-27 rules.

A.2 EASA FRAMEWORK: SC-VTOL AND INNOVATIVE AIR MOBILITY

EASA's *Special Condition for VTOL (SC-VTOL)*, first issued in 2019 and refined through Amendment 2 (2024), defines airworthiness objectives based on operational risk levels (OSL 1–3). The framework differentiates between *enhanced* and *basic* categories, requiring demonstrated equivalent safety to small transport airplanes for enhanced types intended for passenger operations in congested environments.

Complementing SC-VTOL, the Innovative Air Mobility (IAM) Implementation Framework expands regulatory scope to encompass:

- **Vertiport design specifications** (lighting, obstacle limitation, firefighting, and energy interfaces);
- **U-space integration** under *Commission Implementing Regulation (EU) 2021/664*; and
- **Cybersecurity and AI oversight**, aligned with the *EASA AI Roadmap 2.0 (2024)*.

EASA's approach reflects a systems-of-systems philosophy, certifying the ecosystem—aircraft, operations, and digital infrastructure—rather than only the airframe.

A.3 FAA FRAMEWORK: TYPE CERTIFICATION AND OPERATIONAL APPROVALS

The FAA (2023) restructured its regulatory roadmap through the *Innovate28 Plan*, coordinating type certification, pilot training, and airspace integration. eVTOL developers are certified under a *special class process* pursuant to *Title 14 CFR § 21.17(b)*, which allows tailoring of requirements drawn from Parts 23, 27, 29, 33, 35, and 36.

Parallel tracks include:

- **Part 135 Air Carrier Certification**, providing early operational experience with conventional aircraft (e.g., Joby Aviation, 2022);
- **Part 91 Operational Validation Flights** under experimental certificates; and
- **Advisory Circular AC 119–1 (2023)**, integrating cybersecurity protection into safety assessments.

A.4 COMPARATIVE CERTIFICATION MATRIX

Domain	EASA SC-VTOL (2024)	FAA Special Class (2023)	Convergence Potential
Airworthiness basis	Risk-based categories (OSL 1–3)	Case-specific design criteria (§ 21.17 b)	ICAO AAM Study Group Matrix 2025
Propulsion	Electric & hybrid permitted	Electric/turbine hybrid	Shared reliability goals (10^{-9}/hr)
Software assurance	DO-178C/EASA AMC 20–189	DO-178C + FAA Order 8110.49	Mutual recognition MoU
AI functions	Trustworthiness Framework (ethics + explainability)	Concept of “machine learning Level 1–3” validation	ICAO AI Safety Blueprint (2025)
Vertiports	CS-VTOL Subpart E	AC 150/5390-4 Draft	Joint Working Group 2026

A.5 TOWARD GLOBAL ALIGNMENT

Both agencies participate in ICAO’s AAM Study Group (2025), which aims to publish a *Global Airworthiness Alignment Matrix*. Convergence will rest on data transparency and digital evidence exchange—particularly AI-system logs and battery performance datasets. Regulators must also coordinate environmental validation (Annex 16, Volume I Noise, and Volume II Emissions) to ensure consistent sustainability metrics.

REFERENCES (APPENDIX A)

EASA. (2024). *Artificial Intelligence roadmap 2.0*. European Union Aviation Safety Agency. https://doi.org/10.2822/103978

FAA. (2023). *Advanced Air Mobility (AAM) Implementation plan—Innovate28*. Federal Aviation Administration. https://www.faa.gov/AAM-I28

Appendix B: Competency Frameworks and AI-Driven Training Models

B.1 RATIONALE FOR COMPETENCY-BASED TRAINING

The transition to AAM requires a paradigm shift from hours-based licensing to competency-based assessment. Both IATA's Evidence-Based Training (EBT) and Competency-Based Training and Assessment (CBTA) frameworks provide the scaffolding for such evolution. They align well with the EASA Basic Regulation (2018/1139) and the FAA's Integrated Pilot Proficiency Program (2023), emphasizing observable behaviors and risk-management reasoning rather than prescriptive task checklists.

B.2 THE IATA EBT-CBTA MATRIX

IATA's EBT (2023) identifies nine core competencies for flight crews—application of procedures, communication, situational awareness, workload management, problem-solving, decision-making, leadership, automation management, and resilience. Within CBTA, these map to performance indicators measurable through simulation analytics.

Competency Domain	AAM Application Example	Assessment Tool/AI Integration
Automation Management	Supervision of AI flight-control decisions	Explainable AI dashboards; adaptive simulation scoring
Workload Management	Managing multimodal HMI inputs in eVTOL	Eye-tracking and heart-rate variability sensors
Decision-Making	Route and energy optimization under UTM constraints	AI-driven scenario branching (Monte Carlo simulations)
Communication & Teamwork	Multi-pilot or remote support coordination	Virtual reality crew training; speech analytics
Leadership & Resilience	Responding to automation failures	Cognitive stress modeling; biofeedback loops

AI-enabled simulators can deliver personalized competency development by adjusting scenario complexity in real time. The UK CAA Future Flight Challenge (2024) integrates similar adaptive learning loops into its pilot-training pilots.

B.3 HUMAN-AI TEAMING AND COGNITIVE ANALYTICS

AI's role in training extends beyond simulation to learning analytics—using big-data techniques to track skill decay and retraining needs. Ethical use of such analytics requires transparency: trainees must know how data inform evaluation, consistent with EASA's Ethical AI Guidelines (2025).

AAM training centers should incorporate AI ethics modules, ensuring that future eVTOL operators understand algorithmic limitations and human-oversight principles.

REFERENCES (APPENDIX B)

EASA. (2025). *Ethical AI guidelines for trustworthy Artificial Intelligence*. European Union Aviation Safety Agency. https://doi.org/10.2822/700223

FAA. (2023). *Integrated pilot training demonstration*. Federal Aviation Administration.

IATA. (2023). *Evidence-Based Training and Competency-Based Training and Assessment (EBT/CBTA) framework*. International Air Transport Association. https://doi.org/10.47468/iata.ebt.2023

UK RI. (2024). *Future flight challenge phase III roadmap*. UK Research and Innovation. https://www.ukri.org

Appendix C: ROI Models, Economic Forecasts, and Infrastructure Development Strategies

C.1 FRAMING AAM ECONOMICS

Economic sustainability underpins societal legitimacy. Recent analyses (OECD, 2024; World Economic Forum, 2025) indicate a *USD 1 trillion global AAM market by 2040*, contingent on infrastructure coordination and public acceptance. ROI in AAM is a composite of *capital efficiency, operational utilization*, and *regulatory predictability.*

C.2 ECONOMIC MATRIX FOR AAM ROI ASSESSMENT

Input Dimension	Indicators (Examples)	Economic Effect
Capital Investment	Aircraft R&D cost, vertiport construction, grid connection	Determines break-even time
Utilization	Flight cycles per day, load factor, turnaround time	Affects variable cost absorption
Operating Cost	Energy per km, battery lease, maintenance	Influences fare competitiveness
Regulatory Predictability	Certification lead time, airspace access	Reduces risk discount rates
Societal Acceptance	NIMBY index, noise approval rate	Drives route permit success
Sustainability Credits	SAF or renewable energy subsidies	Enhances EBITDA margin

C.3 INFRASTRUCTURE AND ENERGY STRATEGIES

Infrastructure costs represent up to 40% of total AAM capital expenditure. Joint-use vertiports integrated into existing airports can reduce duplication and facilitate multimodal links. EASA's Vertiport Design Manual (2024) and the FAA's Vertiport Engineering Brief No. 1050.1F (2023) emphasize *modular, grid-connected, and community-scaled* sites.

Energy strategy aligns with national Net-Zero 2050 pathways: integrating renewable generation, smart charging, and vehicle-to-grid capabilities. The DOE (2024) initiative on domestic manufacturing for electric aviation components exemplifies how industrial policy can enhance ROI by stabilizing supply chains.

C.4 FORECAST SCENARIOS (2025–2040)

Baseline models forecast compound annual growth rates (CAGR) of *15–18%* through 2040, concentrated in airport shuttles *and* regional mobility. Sensitivity analysis shows public acceptance (+10%) or energy cost (–10%) swings can alter ROI by ± 6–8%, illustrating that *trust and technology are joint economic variables.*

REFERENCES (APPENDIX C)

DOE. (2024). *Domestic manufacturing for electric aviation components initiative*. U.S. Department of Energy. https://doi.org/10.2172/doe.ea.2024

EASA. (2024). *Vertiport design specifications for Innovative Air Mobility*. European Union Aviation Safety Agency.

FAA. (2023). *Vertiport engineering brief 1050.1F*. Federal Aviation Administration.

OECD. (2024). *Urban mobility outlook 2024: Economic and environmental transitions*. OECD Publishing. https://doi.org/10.1787/umo-2024-en

World Economic Forum. (2025). *Global aviation sustainability outlook 2025*. World Economic Forum.

Appendix D: Future Research Directions and Sustainability Policy Recommendations (2025–2050)

D.1 LONG-TERM RESEARCH PRIORITIES

The horizon from 2025 to 2050 will define whether AAM fulfills its environmental promise. Priority research domains include:

1. Hybrid-Energy and Hydrogen Propulsion Safety. Investigations into cryogenic hydrogen storage, fire dynamics, and certification protocols under EASA CS-25H should parallel battery research (EASA, 2024).
2. AI Safety Verification and Explainability. Ongoing development of EASA's "Level-of-Trust Methodology" and NASA's Assured Autonomy Program must converge toward formal verification techniques (NASA, 2024).
3. Social Impact and Equity Studies. Multidisciplinary projects should analyze AAM's effects on urban segregation and accessibility, in line with UN SDG 11 (Sustainable Cities and Communities, United Nations, 2023).

D.2 SUSTAINABILITY POLICY PATHWAYS

Policy frameworks should integrate carbon, noise, and equity indices into performance reviews. By 2035, EASA and FAA could implement "Green Certification Labels" scoring aircraft and operators on life cycle impact. Between 2040 and 2050, ICAO should standardize "Global AAM Sustainability Accounting Standards (G-AAMSAS)" linking operations to net-zero targets.

D.3 GOVERNANCE AND PUBLIC ENGAGEMENT

Long-term trust will depend on co-governance models. Municipal AAM Councils, similar to noise abatement committees, can institutionalize public input on routes, pricing, and data use. Embedding citizen observers in certification sandboxes will translate ethical transparency into procedural practice (OECD, 2024).

D.4 THE 2050 VISION

By 2050, Advanced Air Mobility should constitute an emission-neutral, cyber-resilient, and ethically governed public transport layer integrated with high-speed rail and autonomous ground systems. The maturity of AAM will be measured not by fleet size but by its contribution to human well-being and environmental restoration.

REFERENCES (APPENDIX D)

EASA. (2024). *Artificial Intelligence roadmap 2.0*. European Union Aviation Safety Agency. https://doi.org/10.2822/103978

NASA. (2024). *Advanced Air Mobility mission integration plan*. National Aeronautics and Space Administration. https://doi.org/10.25967/NASA.AAM.2024

OECD. (2024). *Urban mobility outlook 2024*. OECD Publishing. https://doi.org/10.1787/umo-2024-en

United Nations. (2023). *Sustainable development goal 11: Make cities and human settlements inclusive, safe, resilient, and sustainable*. United Nations. https://sdgs.un.org/goals/goal11

Index

Note: Page locators in *italic* indicate a figure, and page locators in **bold** indicate a table on the corresponding page.

For Product Safety Concerns and Information please contact our EU
representative GPSR@taylorandfrancis.com
Taylor & Francis Verlag GmbH, Kaufingerstraße 24, 80331 München, Germany

www.ingramcontent.com/pod-product-compliance
Ingram Content Group UK Ltd.
Pitfield, Milton Keynes, MK11 3LW, UK
UKHW020159200926
475702UK00004B/540
* 9 7 8 1 0 4 1 1 5 5 6 2 1 *